International Management of Tuna, Porpoise, and Billfish

Biological, Legal, and Political Aspects

International Management of Tuna, Porpoise, and Billfish

Biological, Legal, and Political Aspects

by

James Joseph and Joseph W. Greenough

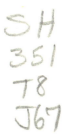

UNIVERSITY OF WASHINGTON PRESS

Seattle and London

Library of Congress Cataloging in Publication Data

Joseph, James, 1930–
   International management of tuna, porpoise, and billfish.

   Includes index.
   1. Fishery management, International. 2. Tuna
fisheries. 3. Porpoises. 4. Billfish fisheries.
I. Greenough, Joseph W., 1936–   joint author.
II. Title.
SH351.T8J67   333.9′5   77–15192
ISBN 0–295–95591–0

Preface

This book is based largely on an earlier document which was prepared by the authors at the request of the Inter-American Tropical Tuna Commission (IATTC). At the IATTC's annual meeting held in October 1976, this original version was circulated to the delegations of the attending nations under the title "Alternatives for International Management of Tuna Resources." In preparing the present volume, all of the material in the original version has undergone moderate revision and updating. A considerable amount of new material has also been added on the tuna-porpoise problem and on billfish management, both of which are considered in the next to last chapter.

All numerical data contained in the text, tables, and figures are from IATTC sources unless otherwise stated. It should be noted that when the original document was being written, 1974 was the last year for which complete data were available on catches, fleet composition, and so forth. Hence, in some cases the data summaries and numerical examples developed to illustrate the consequences of various management systems are based on conditions through 1974. In other instances, where updating seemed appropriate, data from the 1975–77 period have been incorporated in the text.

Although the original version of this book was written at the request of the IATTC, it must be emphasized that it does not necessarily represent the views of, nor does it dictate policy for, the IATTC. Rather this book is a compilation of background material intended to provide a basis for discussions that, hopefully, may lead to solutions to some of the world tuna fishery problems. A number of individuals have reviewed our work in manuscript form and have offered helpful and constructive criticism. We are indebted to all of them and would like to especially acknowledge the valuable suggestions received from Mr.

Gordon Broadhead, Dr. Douglas Chapman, and Dr. Virginia Flagg. All views expressed and conclusions reached, however, are solely our own.

JJ
JWG
La Jolla
February 1978

Contents

Abbreviations   XV

1. **Introduction**   3

2. **The Resources and Their Fisheries**   5

3. **Existing International Arrangements for Tuna Management**   13
   Inter-American Tropical Tuna Commission   13
   International Commission for the Conservation of Atlantic Tunas   18
   Indian Ocean Fishery Commission and Indo-Pacific Fisheries Council 20

4. **Major Problem Areas**   22
   Collection and Analysis of Data for Management of the Fisheries   23
   Distribution of the Catch among Harvesters   25
   Economics and Fleet Carrying Capacity   32
   Enforcement of Conservation Regulations   35

5. **Conservation and Tuna Management Philosophy**   40

6. **Control to 200 Miles by Individual Coastal States**   43
   Exclusive National Fishing Zones   44
   Licensing by Individual Nations   48

7. **Extension of the Present Eastern Pacific Overall Quota System** 53

 8. **Open Access with Participant Fees and Resource Adjacency Allocations (PAQ Management)**   58
    Partial Catch Allocation and Resource Adjacency   60
    Open Access Fishing and International Licensing   67
    Collection and Redistribution of Participant Fees   73
    Other Aspects of PAQ Management   85

 9. **Regional Coalitions**   93
    RAN Coalitions with Exclusion of Non-RANs   94
    RAN Coalitions with Licensing of Non-RANs   100
    Coalitions of RANs and Non-RANs with Allocations   107
    Coalitions of RANs and Non-RANs without Allocations   114

10. **Total Allocation of the Resource**   120

11. **Resource Allocation by Competitive Bidding**   131

12. **The Special Problems of Porpoise and Billfish Conservation**   136
    The Fishery on Associated Schools of Tuna and Porpoise   137
    The MMPA and Ensuing Regulations   144
    Development of Fishing Technology to Reduce Porpoise Mortality   154
    Internationalization of the Tuna-Porpoise Problem   168
    Billfish Conflicts Involving Commercial and Sport Fishermen   172

13. **Institutional Arrangements for Management of Tuna and Associated Species**   180
    Need for Coordinated Global Management   180
    Functions and Division of Responsibility   183
    Staffing   187
    Membership and Voting   190
    Management at the Regional Level and Species Coverage   193
    Toward a Global Management System   195

    Appendix I. Achievement of Quotas and Fishery Closure   198

Appendix II. Differential Treatment of Yellowfin and Skipjack under PAQ Management   202

Appendix III. Changes in Fleet Composition: Development of RAN Fleets   206

Pros and Cons of Flag Changes Versus New Construction   206

Effect of Transition to PAQ Management on RAN and Non-RAN Fleets   212

Factors Mitigating the Impact of PAQ Management   222

Appendix IV. Control of Entry   229

An Illustrative Licensing System for Control of Fleet Growth   229

Other Problems Relating to Control of Fleet Growth   237

Index   243

Figures

1. World catches of tuna and billfish by species and market groups in 1974   6
2. Catches of principal market species of tuna by oceans in 1974   7
3. Bait fishing   8
4. Longlining   9
5. Purse seining   10
6. Trends in fleet carrying capacity, annual catch per ton of carrying capacity, catch value, and earnings per ton of carrying capacity for the eastern Pacific tuna fishery, 1963–75   16
7. Approximate boundaries of areas extending 200 miles from shores of coastal nations in the eastern Pacific   28
8. Approximate range of the offshore spotted porpoise stock   138
9. Approximate ranges of the eastern and whitebelly spinner porpoise stocks   140
10. NMFS interpretation of the optimum sustainable population concept as set forth in the MMPA   148
11. Design of the super apron system tested by the *Elizabeth C.J.*   164
12. Billfish migrations as determined through tagging programs undertaken by various sportsmen groups and agencies   175
13. Four alternatives for phasing in combined national yellowfin allocations in the eastern Pacific   214
14. Changes in RAN and non-RAN fleet carrying capacities, catches, and catches per capacity ton over a 10-year allocation phase-in period   218

Tables

1. Catches of principal market species of tuna by nations in 1974 11
2. Annual yellowfin and skipjack catches taken within 200 miles of coastal nations and beyond 200 miles in the eastern Pacific by the entire international fleet 30
3. Maximal and minimal criteria for partially allocating an overall yellowfin quota in the eastern Pacific 64
4. Annual yellowfin and skipjack catches taken within 12 miles of coastal nations in the eastern Pacific by the entire international fleet 70
5. Examples illustrating collection and redistribution of participant fees under: (1) 1970–74 catch distribution; (2) high allocations; (3) low allocations 80
6. Simulated annual catches by RAN and non-RAN fleets 99
7. Average monthly catch rates for yellowfin and skipjack in national 200-mile zones and beyond 200 miles by purse seiners over 400 tons, 1970–74 116
8. Estimated porpoise kills for principal stocks by vessels of all flags in the eastern Pacific Ocean, 1959–77 143
9. Condition of porpoise stocks in 1976 as compared to their unexploited sizes 151
10. Annual United States porpoise kill quotas by stock, 1978–80 153
11. Summary of porpoise-saving concepts developed by NMFS and United States tuna fishing industry 158
12. Billfish catches by commercial fishing operations in 1975 by all nations combined 176
13. 3 possible methods for phasing in national yellowfin allocations in the eastern Pacific over a 10-year period 210
14. Definition of 16 hypothetical cases examining changes in

RAN and non-RAN fleet carrying capacities, total catches, and catches per capacity ton over a 10-year allocation phase-in period   216

15. Final RAN and non-RAN fleet carrying capacities, total catches, and catches per capacity ton for all 16 allocation phase-in cases of Table 14   220

16. Possible significance of factors tending to reduce the impact of PAQ management with RAN yellowfin and skipjack allocations set at two-thirds of their high levels   226

17. Possible significance of factors tending to reduce the impact of PAQ management with RAN yellowfin allocations set at their high levels and skipjack unallocated   227

18. 33 possible types of changes, grouped into 10 categories (A-J), in regional URAN, DRAN, and non-RAN fleets   232

19. Effects of 10 types of fleet change (A-J) on URAN, DRAN, and non-RAN fleets and on associated licensing transactions   234

Abbreviations

CYRA	Commission's Yellowfin Regulatory Area
IATTC	Inter-American Tropical Tuna Commission
ICCAT	International Commission for the Conservation of Atlantic Tunas
ICNAF	International Commission for Northwest Atlantic Fisheries
IOFC	Indian Ocean Fishery Commission
IPFC	Indo-Pacific Fisheries Council
IWC	International Whaling Commission
MMPA	Marine Mammals Protection Act
MNP	Maximum net productivity
MSP	Maximum sustainable population
NMFS	National Marine Fisheries Service
NPFSC	North Pacific Fur Seal Commission
OSP	Optimum sustainable population
PAQ	Partially allocated quota
RAN	Resource adjacent nation

International Management of Tuna, Porpoise, and Billfish

Biological, Legal, and Political Aspects

1 Introduction

The stocks of tuna and tuna-like fishes support major commercial and sports fisheries in temperate and tropical oceans throughout the world. In terms of product value, tuna constitute one of the most important living resources of the sea. As is true for any renewable marine resource, there is a limit to what tuna stocks are capable of producing on a sustainable basis, and this level of sustainable productivity is related to the intensity with which the stocks are exploited. Consumption of tuna has increased steadily over the last three decades, being limited mainly by production in recent years. This strong demand coupled with a limited supply is generating strong competition both within and among nations to harvest the available fish. The increasing exploitation of tuna stocks makes development and implementation of sound conservation programs imperative. These programs should be designed to maintain the harvest of tunas at near-maximum sustainable levels, to protect stocks from overexploitation, and to maintain viable tuna industries. Accomplishing all of this without serious political confrontations will require that every effort be made to take account of the major economic, political, and social factors operating within the fisheries and among nations participating in them.

Tuna possess unique qualities that necessitate special consideration if they are to be managed properly: they are highly migratory, making vast transoceanic migrations during short periods of time; they may be in the coastal waters of one nation today and in those of another nation or on the high seas tomorrow; and their migration routes, areas of abundance, and availability vary significantly from year to year. In short, they recognize no man-made boundaries, being subject only to the dynamic forces of their environment. Tuna are sought by fishermen using small coastal canoes operating in very limited areas as well as by fleets of large oceanic vessels that may fish three oceans in a single year. Catches are bought and sold in an international market, with the raw product being landed, transshipped, and processed in many ports throughout the world. When these unique characteristics are taken into account, it is clear that development of a system of management that will result in adequate conservation of tuna resources will dictate a broad international approach.

Several regional fisheries bodies deal with the scientific study and management of tuna. Some have been fairly successful in addressing research and management problems; others have accomplished little of significance. None, however, has yet instituted a management program capable of responding to the problems looming in the future, especially those closely associated with the nearly universal trend toward extended fisheries jurisdiction. Extension of fisheries jurisdiction to 200 miles would place many of the world's most productive tuna fishing areas within the province of coastal nations presently without the capability to harvest the resources fully. Many of these nations do not feel that management controls should hinder them from developing their tuna fishing industries. Conversely, nations with well-developed tuna industries do not want to see their catch shares diminished, and they want their fleets to be able to fish wherever they wish. Obviously the goals of the two groups of nations are conflicting.

The basic purpose of this inquiry is to explore alternative approaches to resolving the conflicts just described, for, if they cannot be resolved, economic havoc and overexploitation of the tuna and tuna-like resources seem likely. First, we will discuss the unique qualities that set these species apart from other commercially important stocks of fish, and identify certain basic problems that must be addressed in establishing an effective management program. Then we will consider a number of alternative management systems and institutional arrangements. Because there is a long history of tuna management in the eastern Pacific Ocean, and because an extensive data base exists, the discussion of alternative management systems will concentrate primarily on that area.

Finally, we will review two management problems involving species often captured in association with tuna. The first problem is unique to the eastern Pacific and involves the mortality of porpoise with which yellowfin tuna are often associated and captured. The second involves conflicts between sports fishing interests and commercial fishing interests over utilization of billfish resources.

2 The Resources and Their Fisheries

The tuna and tuna-like fish include a large number of diverse species that for management purposes are generally considered together because they frequently are taken in the same fishing operations. They are most often grouped into three categories (Fig. 1). The first category includes six major tuna species, making up about 75 percent of the world catch of tuna and tuna-like fishes, which are called the principal market species: yellowfin *(Thunnus albacares)*, bigeye *(T. obesus)*, albacore *(T. alalunga)*, northern bluefin *(T. thynnus)*, southern bluefin *(T. maccoyii)*, and skipjack *(Katsuwonus pelamis)*. The second category, making up about 22 percent of the world catch, consists of species commonly referred to as secondary market species. These are generally smaller and less heavily exploited species, and include among others bonito *(Sarda* spp.), black skipjack *(Euthynnus* spp.), and frigate mackerel *(Auxis* spp). The third category, the billfish, includes all members of the families Xiphiidae and Istiophoridae; they account for about 3 percent of the world catch of tuna and tuna-like species.

Most tuna and tuna-like species are highly mobile and in many instances undertake extensive migrations. In the Pacific Ocean, northern bluefin and albacore migrate between the nearshore waters off Canada, Mexico, and the United States and Japanese waters. Skipjack also migrate extensively, traveling at least between the central Pacific and the coastal waters of both the eastern Pacific and Japan. In the Atlantic Ocean northern bluefin travel between the Gulf Stream waters off North America and European waters from Spain to Norway, and albacore probably migrate just as extensively. Southern bluefin, found only in the southern hemisphere, migrate from spawning areas around Australia to the Atlantic, Pacific, and Indian oceans. Yellowfin and bigeye, although undertaking migrations of several thousand miles, do not appear to make such extensive migrations as the

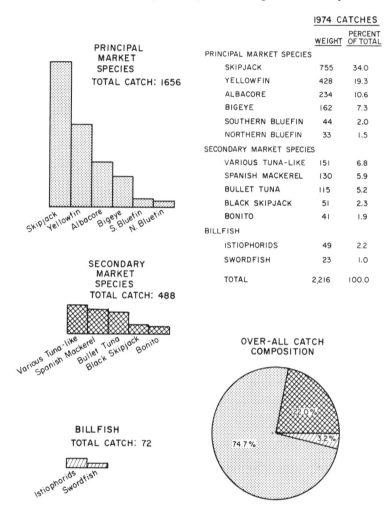

1974 CATCHES		
PRINCIPAL MARKET SPECIES	WEIGHT	PERCENT OF TOTAL
SKIPJACK	755	34.0
YELLOWFIN	428	19.3
ALBACORE	234	10.6
BIGEYE	162	7.3
SOUTHERN BLUEFIN	44	2.0
NORTHERN BLUEFIN	33	1.5
SECONDARY MARKET SPECIES		
VARIOUS TUNA-LIKE	151	6.8
SPANISH MACKEREL	130	5.9
BULLET TUNA	115	5.2
BLACK SKIPJACK	51	2.3
BONITO	41	1.9
BILLFISH		
ISTIOPHORIDS	49	2.2
SWORDFISH	23	1.0
TOTAL	2,216	100.0

Figure 1. World catches of tuna and billfish by species and market groups in 1974 (in thousands of metric tons)

other principal market species. Many of the secondary market species, although available information is scanty, appear to be less migratory than the principal market species. Billfish are quite migratory, with some species making migrations of several thousand miles.

Prior to 1940, production of the principal market species of tuna never exceeded 300,000 metric tons per year. After World War II

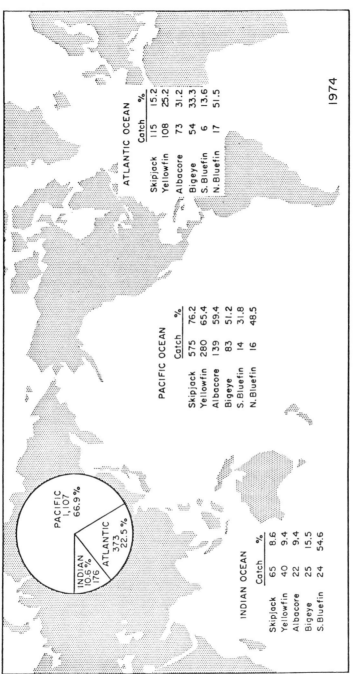

PACIFIC OCEAN	Catch	%
Skipjack	575	76.2
Yellowfin	280	65.4
Albacore	139	59.4
Bigeye	83	51.2
S. Bluefin	14	31.8
N. Bluefin	16	48.5

ATLANTIC OCEAN	Catch	%
Skipjack	115	15.2
Yellowfin	108	25.2
Albacore	73	31.2
Bigeye	54	33.3
S. Bluefin	6	13.6
N. Bluefin	17	51.5

INDIAN OCEAN	Catch	%
Skipjack	65	8.6
Yellowfin	40	9.4
Albacore	22	9.4
Bigeye	25	15.5
S. Bluefin	24	54.6

PACIFIC 1,107 66.9%

ATLANTIC 373 22.5%

INDIAN 10.6% 176

1974

Figure 2. Catches of principal market species of tuna by oceans in 1974. Catches of each species in each ocean are expressed in terms of thousands of metric tons and as a percentage of the total world catch of that species. Comparable figures for all principal market species combined are given in the pie diagram

annual production began to increase rapidly, exceeding one-half million tons by the early 1950s and reaching a million tons by the mid-1960s. Since then annual production has increased further, reaching 1.7 million tons in 1974. Skipjack account for much of this recent

Figure 3. Bait fishing. A typical eastern Pacific bait boat of about 75 tons is shown. The fishermen in the racks are catching tuna on barbless feathered jigs. The livebait tank is behind them. The chummers beside the tank are throwing the bait to attract and excite the tuna into a feeding frenzy

increase. Of the principal market species, skipjack is the most important in terms of tonnage harvested, making up 46 percent of the world catch. It is followed by yellowfin (26 percent), albacore (14 percent), bigeye (10 percent), southern bluefin (3 percent), and northern bluefin (2 percent). The largest catches are made in the Pacific Ocean (67 percent), with the Atlantic Ocean (22 percent) and the Indian Ocean (11 percent) following (Fig. 2).

Three basic forms of fishing account for nearly all the commercial catch of principal market species. Bait fishing (Fig. 3) is presently the most important method, followed by longlining (Fig. 4) and purse seining (Fig. 5). Bait fishing and purse seining are both surface fishing methods, but purse seining is much more effective in terms of catch per unit of effort expended.

Because tuna and tuna-like fish occur off the coasts of nearly all nations bordering on tropical and temperate seas, a large number of nations participate in tuna fishing. During 1974 about forty nations reported capturing commercial quantities of tuna, but just six nations

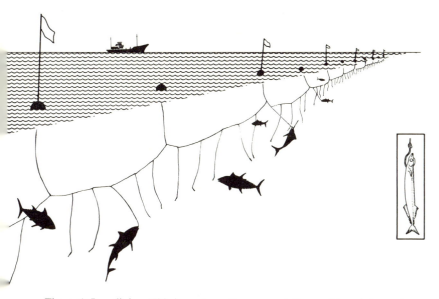

Figure 4. Longlining. This is a schematic representation and is not drawn to scale. In a typical longlining operation about 2,200 hooks might be fished from an 80-mile-long main line supported by floats at 300-meter intervals. The insert shows saury being used for bait. Squid is another common bait

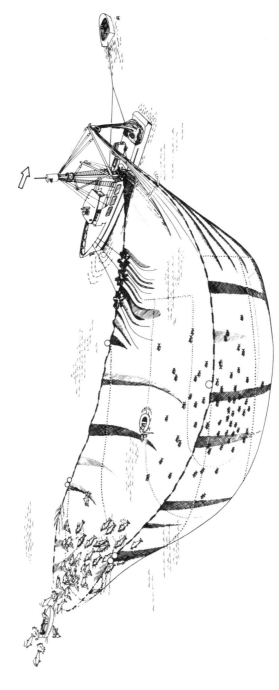

Figure 5. Purse seining. The purse seiner has made a set on a school of yellowfin tuna associated with a porpoise school. The seine has been pursed shut and partially hauled aboard over the hydraulic block. The seiner is depicted in the process of backing down to release the school of porpoise. A fine mesh safety panel and attached apron surround the release area. (See Chap. 12 for a further discussion of the porpoise release techniques illustrated here.) (Source: NMFS)

took nearly 77 percent of the principal market species catch (Table 1). The two most dominant nations in terms of both production and consumption are Japan and the United States, who take about 55 percent and utilize about 75 percent of the total world catch of the principal market species. Japan catches about 40 percent and consumes about 30 percent, whereas the United States catches about 15 percent and consumes about 45 percent. The nations of western Europe consume about 20 percent of the catch, with the remaining 5 percent being utilized throughout the rest of the world.

The carrying capacity of the world tuna fleet has been steadily increasing for the last fifteen years. In the early 1960s total capacity was about 350,000 metric tons. By the end of 1972 capacity had more than doubled, and it continues to increase. Fleet carrying capacity may already exceed that needed to harvest the present world catch. This seemingly excessive fleet growth creates major problems in the management of tuna resources.

Since there is such a great demand for tuna, it is of interest to explore possibilities for increasing production. Except for skipjack, all major stocks of the principal market species appear to be fully or almost fully exploited, and some stocks may be overexploited. Yellowfin, albacore, and bigeye tuna appear to be nearly fully exploited, throughout the world. Increased effort on these species would result

TABLE 1

CATCHES OF PRINCIPAL MARKET SPECIES OF TUNA BY NATIONS IN
1974

(Thousands of metric tons)

Country	Principal Market Species Catch	Percent of Total
Japan	649	39.2
U.S.A.	258	15.6
Republic of Korea	104	6.3
Spain	93	5.6
Republic of China	91	5.5
France	76	4.6
Subtotal	1,271	76.8
Other nations	385	23.2
Total	1,656	100.0

at best in only small catch increases and could possibly cause decreased catches. Northern bluefin are probably fully exploited in the Pacific and may be overexploited in the Atlantic. Southern bluefin are heavily exploited and catches have recently declined by nearly 30 percent. Even though skipjack is already the most important species in the world tuna catch, it appears, judging from biological data, that they are still underexploited throughout much of their range. Thus, skipjack production could probably be increased. An exception may be in parts of the western Pacific, where some Japanese studies suggest that the skipjack catch may be approaching an upper sustainable limit.

Little work has been done on assessing stocks of secondary market species and billfishes, and such work is urgently needed. It does appear, however, that catches of many secondary market species could be substantially increased if certain marketing and technological impediments could be overcome. Many species of billfish seem to be fully exploited and, in some cases, perhaps overexploited.

If the world is to maintain tuna resources at population levels that can sustain near-maximum physical yields, it is clear that controls on exploitation will be required. If the objective is to optimize something other than physical yield (i.e., social, economic, or recreational benefits), then even more stringent controls will be needed. Without some system for determining and implementing the necessary controls, the major stocks of tuna are likely to be overexploited and returns to society diminished. But implementation of effective management controls will require resolution of the fundamental differences between those nations presently involved in the tuna fisheries and those hoping to develop such fisheries.

3   Existing International Arrangements for Tuna Management

There are four major international organizations concerned with the scientific study and management of tuna: the Inter-American Tropical Tuna Commission (IATTC), the International Commission for the Conservation of Atlantic Tunas (ICCAT), the Indian Ocean Fishery Commission (IOFC), and the Indo-Pacific Fisheries Council (IPFC). The IATTC and ICCAT are concerned with tuna, tuna-like species, and tuna baitfish. The IATTC is also concerned with several species of porpoise that are captured in association with tuna. The IOFC and the IPFC are responsible for all species within their geographical areas, including tuna.

Inter-American Tropical Tuna Commission (IATTC)

A major fishery for yellowfin and skipjack has operated in the eastern Pacific Ocean since the end of World War II. Although the fishery in the area was originally developed as a bait fishery, purse seining is now the dominant harvesting method. The United States is by far the most important harvesting nation, but several Latin American nations are steadily developing their fleets. Several nations from outside the eastern Pacific area also participate in the fishery. Because of concern over tuna and bait-fish resources, Costa Rica and the United States entered into an agreement in 1949 to establish the IATTC, with its own internationally recruited scientific staff. The convention establishing the IATTC is open to adherence by other nations that fish in the eastern Pacific, and as of 1977 there were eight member nations: Canada, Costa Rica, France, Japan, Mexico, Nicaragua, Panama, and the United States. The duties of the IATTC are to study tuna and other fish caught by tuna fishing vessels within its geographical area of responsibility and, if necessary, to recommend management measures designed to maintain stocks at levels that will produce maximum yields on a sustained basis. In 1966 IATTC studies led to the establishment of an overall annual catch quota for yellowfin to be taken by vessels of all nations within a specified area of the eastern Pacific known as the Commission's Yellowfin Regulatory Area (CYRA). Initially the CYRA encompassed the entire eastern Pacific yellowfin fishery, but in recent years significant catches have been taken west of the CYRA. Within the CYRA the quota is taken on a first-come, first-served basis, implying that the resources belong to whoever can first catch them. This overall quota for which fleets of all nations in the world can compete has sometimes been referred to as a global quota. The quota has been modified from time to time as changing conditions have warranted and, generally speaking, the management program for the conservation of yellowfin has operated effectively since 1966 in terms of maintaining populations at desirable levels of abundance. The management program, however, faces a number of serious problems, and, unless these can be resolved, continued success will be extremely difficult if not impossible.

Many of the difficulties threatening maintenance of a viable management program in the eastern Pacific revolve around the distribution of the allowable catch among nations. When the quota is reached and the yellowfin season in the CYRA closes, it is obvious that the largest shares of the allowable catch will have been taken by the nations with the largest fleets. However, developing Latin American nations adjacent to waters that produce large shares of the catch are now entering the tuna fisheries or have aspirations to do so. Most consider the traditional common property, open-access concept outdated because it does not recognize what they consider to be their special rights to shares of a resource that spends a part of its life in their coastal waters. In their view it is irrational and inequitable for their fleets to have to compete on equal footing with fleets of developed nations, especially the very large United States fleet. Therefore, they strongly favor allocating a portion of the allowable catch among themselves on the basis of adjacency to the resource, with the remainder being distributed in some way among harvesting nations.

In opposition to the Latin American position, developed nations, particularly the United States, have strongly defended the common property, open-access philosophy and the importance of historic fishing rights. They maintain that tuna, by virtue of their highly migratory nature, are not the property of any particular nation, but belong to whoever can catch them. Also, it was United States expertise and investment that developed the eastern Pacific fishery. For years the United States took nearly 100 percent of the catch, but, since the implementation of the yellowfin management program in 1966, their share of the combined yellowfin and skipjack catch has fallen to about 65 percent. In their view any further special allocations or preferential treatment for developing coastal states comes from what should be their fair share of the fishery that they pioneered.

Exacerbating the whole catch distribution problem has been the tremendous fleet growth in recent years. The demand for tuna has been steadily increasing, but increases in supply have not kept pace. This together with the anticipation of profits has resulted in intense competition among countries and among industry elements within countries for the available catch. The result has been an unprecedented increase in fleet carrying capacity. During the mid-1960s the international surface fishing fleet in the eastern Pacific Ocean had a cumulative carrying capacity of about 44,000 short tons, the United States share being a little over 90 percent. At the end of 1975 the international fleet totaled about 169,000 tons (183,000 tons by the end of 1977), with the United States share being about 124,000 tons or roughly 73 percent. This 125,000-ton expansion between the mid-1960s and 1975 constituted an increase of over 280 percent, which far exceeded the 80 percent increase in the catch of the eastern Pacific fleet over the same period. The inevitable result has been a decrease in production of individual vessels. Expressed as catch in short tons per short ton of carrying capacity, a steady decline of from 5.0 tons in 1967 to a low of 2.3 tons in 1975 is evident (Fig. 6). In other words, the number of full trips made by the average vessel decreased from about five in 1967 to less than two and a half in 1975. (These figures include the catch of all species in all areas, including the Atlantic, made by vessels fishing at least part of the year in the eastern Pacific.) These trends go a long way toward explaining the expansion of fishing by the eastern Pacific fleet into areas west of the CYRA and into the Atlantic.

Over the 1967–74 period the value per ton of catch increased steadily. In terms of gross earnings per capacity ton, this increase in value more than offset the decline in catch per ton of capacity through 1971. The situation was reversed in 1972 and 1973 when the loss of catch overshadowed the increase in catch value, but the situation reversed itself again in 1974 due to a large price increase. In 1975 prices fell for the first time in seven years, and gross earnings declined to below

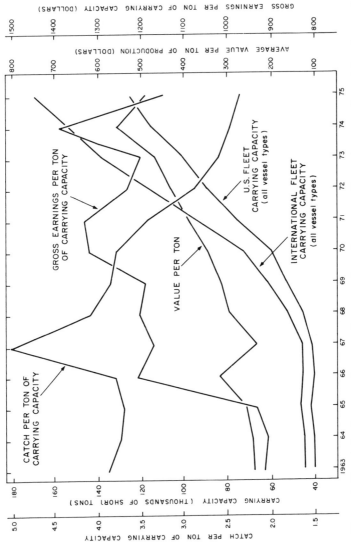

Figure 6. Trends in fleet carrying capacity, annual catch per ton of carrying capacity, catch value, and earnings per ton of carrying capacity for the eastern Pacific tuna fishery, 1963–75

the 1973 level. Thus, gross earnings per ton of carrying capacity rose from 1967 through 1971, declined in 1972 and 1973, rose again in 1974, but then declined sharply in 1975. Owing to escalating costs, net earnings must have performed more poorly than gross earnings over the period.

The economic situation just described poses a very real threat to the conservation program in the eastern Pacific. As fleet capacity is allowed to grow under the pressure of intense competition for supplies of raw tuna, it reaches a size that for economic viability requires harvest levels far exceeding the productive capability of the resource. The result will be political confrontation. Demands for special allocations to insure viable and economically sound industries will be made by both nations and industry user groups. Unless these demands are in some way recognized and dealt with, the ultimate result may well be disregard for conservation measures.

In order for a conservation program to succeed, any management regulations that are adopted must be enforced. Success of the present management system in the eastern Pacific requires that each nation participating in the fishery (both members and nonmembers of the IATTC) assume responsibility for enforcing with regard to its own nationals any IATTC conservation recommendations that are adopted. To do this effectively, a nation must have a capability to locate its vessels at sea on an almost daily basis and also a capability to determine the species of tuna being captured. Unfortunately not all nations fishing tuna in the CYRA are presently both physically able and politically willing to enforce IATTC conservation recommendations.

The United States has domestic legislation requiring that all nations fishing for tuna in the eastern Pacific participate in the conservation program. A 1962 amendment to the Tuna Conventions Act of 1950 prohibits the United States from promulgating regulations for the control of tuna fishing in convention waters unless "all countries whose vessels engage in fishing for species covered by the convention in the regulatory area on a meaningful scale, in terms of effect upon the success of the conservation program, [adopt] effective measures for the implementation of the Commission's recommendations. . . ."
Furthermore, "The Secretary of Commerce shall suspend at any time the application of any such regulation when, after consultation with the Secretary of State and the United States Commissioners, he determines that foreign fishing operations in the regulatory area are such as to constitute a serious threat to the achievement of the objectives

of the Commission's recommendations." In other words, if any other nation fishes on a scale that the United States believes could affect the success of the conservation program but is unable or unwilling to enforce regulations recommended by the commission, then neither will the United States regulate its fishing operation. This is important because the United States is the major producer of tuna in the eastern Pacific, and without its participation and support there can be no effective management program.

International Commission for the Conservation of Atlantic Tunas (ICCAT)

The fishery for tunas in the Atlantic, with a few minor exceptions, began much later than in the eastern Pacific. It was dominated for many years by longliners, but in the late 1950s bait boats and purse seiners quickly became major participants. Presently, Japan, the Republic of China, and the Republic of Korea are the leading longlining nations in the Atlantic, while France (and her former colonies), Japan, Spain, and the United States dominate the purse-seine and bait-boat fisheries. As these fisheries developed, the total catch of the principal market species increased rapidly, causing concern by the mid-1960s over the rate of exploitation. In 1969 the ICCAT was created and given responsibility for the scientific study and management of the tuna and billfish in the Atlantic Ocean. As of 1977 the ICATT member nations were: Angola, Brazil, Canada, Cuba, France, Gabon, Ghana, Ivory Coast, Japan, Republic of Korea, Morocco, Portugal, Senegal, South Africa, Spain, the United States, and the USSR.

The ICCAT is structured differently from the IATTC in that it has not been provided with sufficient funding and staff to conduct its own research. Although efforts have been made in recent years to strengthen the ICCAT budget, the collection of basic catch and effort data and the conducting of biological studies remain the responsibility of member governments. This lack of adequate funding has inhibited data acquisition and the timely formulation of management advice.

Much concern has been expressed by some ICCAT members over the condition of yellowfin and northern bluefin tuna stocks, and minimum size limits have been set for both species. In addition, for northern bluefin, member nations are cooperating to limit fishing mortality to recent levels. No consensus has been reached, however, on the need to control harvest levels through establishment of quotas. The differing opinions as to the need for management reflect differences in scientific opinion that stem partially from a lack of adequate statistical information (catch, effort, and size composition data) upon which to base scientific analysis of the effect of fishing on the stocks. These deficiencies result from the diffusion of authority and responsibility for data collection among participating nations. If at the outset the ICCAT had been assigned responsibility for data collection and had received the necessary fiscal support, the scientific community would now be in a much better position to assess the effects of exploitation on tuna stocks in the Atlantic and to make necessary management recommendations.

In addition to the technical problems of data collection and analysis, problems similar to those evident in the eastern Pacific involving catch distribution and the economics of harvesting are important in the Atlantic. Many developing coastal states adjacent to waters in which tuna are harvested are reluctant to agree to any conservation scheme that involves catch quotas unless special allocations are included recognizing what they believe are their rights as developing coastal states. Unchecked fleet growth is also exacerbating the problem of catch distribution in the Atlantic. The intensity of fishing by the international fleet, principally off west Africa, is growing each year owing to accelerating construction of new vessels. In most years there is also a large seasonal influx of vessels from the eastern Pacific after closure of unrestricted fishing there. As in the Pacific, production increases have not kept pace with fleet growth.

A problem in the Atlantic not encountered in the eastern Pacific involves competition between gear types. The surface fishery (bait boats and purse seiners) harvests young fish which, if allowed to grow, would have become available to the longline fishery. Largely because of this competition, the expansion of the surface fishery in the 1960s was accompanied by about a 40 percent reduction in longline yellowfin catches.

Development of an effective system for enforcing any management regulations recommended by the ICCAT will be just as important in the Atlantic as in the Pacific. At the present time, enforcement of size limits for yellowfin and bluefin is the responsibility of each individual government with respect to its own citizens and vessels. Because of the nature of the fishery and the logistics involved, however, enforcement of these regulations has not been uniform. Obviously some better system for enforcement is needed, especially in the case of catch limits. For adequate management, a capability to inspect catches of individual vessels and to fix the location of vessels on the high seas will be required.

Indian Ocean Fishery Commission (IOFC) and Indo-Pacific Fisheries Council (IPFC)

In the Indian Ocean, principal market species other than skipjack are taken mostly by three nations fishing longline gear: Japan, the Republic of China, and the Republic of Korea. Additionally, coastal fleets using drift nets and hook and line gear take moderately large catches of skipjack in the Indian Ocean. In the western Pacific a primarily Japanese bait fishery takes the largest catches of skipjack made anywhere in the world. Also, Japan, the Republic of China, and the Republic of Korea operate substantial longline fisheries for the other principal market species.

The IOFC was established in 1967 by the Council of the Food and Agricultural Organization (FAO) of the United Nations under Article VI-1 of the FAO constitution. The commission has broad responsibility over the entire field of fishery research and development in the Indian Ocean. The IPFC was formed in 1948, also within the framework of FAO, under Article XIV of the FAO constitution. Like the IOFC, the IPFC has broad responsibility over the development and conservation of the living marine and freshwater resources of the Indo-Pacific region.

Neither body employs a permanent secretariat or staff. Stock assessment studies have been limited to general reviews made by a special *ad hoc* group of experts serving both bodies. The single most important point that this group has made, and it has made it repeatedly, is that basic data on catch, effort, and size composition of the catch needed for proper assessment of western Pacific and Indian Ocean tuna stocks are inadequate in terms of both coverage and timeliness. This group has expressed concern over the condition of certain fisheries in both the western Pacific and Indian oceans, particularly the longline fisheries, and has strongly recommended that the IOFC and IPFC be given the responsibility and support necessary to collect the requisite data. Unless some arrangement for accomplishing this can be made, proper tuna management cannot be attained in these two areas.

In addition to the basic problems of data acquisition and analysis in the Indo-Pacific region, there exist the problems of catch distribution and fleet expansion that are found in other ocean areas. Similarly, if and when conservation recommendations are adopted, effective systems for enforcement will be required. An additional problem unique to these FAO commissions is that non–United Nations members cannot belong. For example the Republic of China, one of the most important tuna fishing nations, cannot be a member of these international bodies. Management of tuna cannot succeed unless all significant participants in the fisheries are full partners in any agreement.

4   Major Problem Areas

From the foregoing discussion it is clear that problems associated with the scientific study and management of tuna are quite similar in all major oceans of the world. These problems can be grouped into four general areas. The first has to do with the study of the animals themselves and involves collection of basic fishery statistics, biological data, and environmental data, with the timely analysis of these data being of prime importance. The second problem area concerns the distribution of the catch among users, be they individuals or nations. The third area involves the economics of fleet operation, particularly with regard to recent rapid increases in fleet carrying capacities. The fourth concerns problems of enforcing any conservation regulations that might be adopted by involved nations.

All of these four problems, but especially the problems of catch distribution and enforcement, are intimately related to the question of who has jurisdiction over fisheries resources in coastal waters. It is likely that some form of extended fisheries jurisdiction to 200 miles will become a reality on a universal basis. This could result either from international agreement within the context of the current Conference on the Law of the Sea or from a series of unilateral actions by coastal states. Indeed, many nations have already taken unilateral action.

Extended jurisdiction can take several forms. For example Ecuador and Peru claim 200-mile territorial seas, while Mexico and Costa Rica claim 12-mile territorial seas and adjacent 188-mile zones of economic influence within which they have jurisdiction over fisheries resources. In 1976, United States legislation created a 197-mile fishery conservation zone extending beyond its 3-mile territorial sea. Within this fishery conservation zone, the United States claims jurisdiction over all coastal species, but it also provides for international management of highly migratory pelagic species that occur in the zone. Obviously the method of catch distribution and the nature of enforcement will be related to whether management is on a national or an international basis.

Collection and Analysis of Data for Management of the Fisheries

The basic requirement for management-oriented research on tunas is to have available catch statistics by species as well as data on the fishing effort that produced these catches. Also of prime importance are samples showing the size composition of the catch and various other kinds of data that provide information on life histories and relationships between the fish and their physical and biological environment. The immediate research goals are to assess condition of stocks and to determine optimum harvest levels in order to establish and maintain an effective management program. A longer-term objective is to achieve a better understanding of all aspects of tuna biology by undertaking studies that are primarily scientifically oriented rather than management oriented. Once a management program has been implemented, it is necessary to monitor fishing activities on virtually a real-time basis in order that management decisions such as when to close the fishery can be made effectively.

On numerous occasions prior to the tremendous expansion of world tuna fleets in recent years, scientists have stressed the need for collecting adequate statistics for management of the fisheries and have predicted the difficulties that would be encountered if they were not collected. In some areas international bodies have been provided with adequate staff and funding for collection of the data required to meet management needs (e.g., the IATTC and to a lesser extent the ICCAT). Elsewhere, as mentioned earlier, collection of basic fishery statistics has been woefully inadequate. For example, in the Indian Ocean estimates of skipjack catches may be in error by more than 50 percent, while for certain areas of the western Pacific it is impossible to make reasonable estimates of the large quantities of tuna that are apparently taken. Estimates of fishing effort in these areas are even less reliable and in many cases are completely lacking. Also, information on the size composition of the catch is generally nonexistent.

From these experiences it is clear that if international management of tuna resources is to be successful, collection of basic statistical data must be carried out by a management organization that is given a mandate to accomplish this job and is provided with the fiscal support necessary to carry it out. Furthermore, it is absolutely necessary that stocks under exploitation be monitored and managed over their entire ranges. This will require coverage of vast geographical areas, because single populations of tuna in some cases extend across entire oceans, and one species, the southern bluefin with its circumpolar distribution, extends throughout all the southern oceans. Management cannot be applied effectively over only a portion of a population's range, because what happens in another portion of its range will affect the entire population.

Because stocks of different species of tuna are frequently overlapping, because the fleets that fish for them are highly mobile, and because the raw product may pass through markets in two or three nations in a matter of months, it seems redundant and inefficient for a number of organizations to gather data independently. Ideally, it would be far more efficient (both economically and logistically) for the responsibility for collection and dissemination of such data to reside in a single international body. Because the collection of basic data and their subsequent analysis are so closely interwoven, it is also reasonable that the same body should be responsible for basic analysis of the data.

It is encouraging to note that most of the data collection problems discussed above are quite tractable. In the eastern Pacific, the IATTC has demonstrated that systems can be developed for monitoring the fishery, analyzing resulting data, and instituting management controls essential for rational utilization of the tuna resources. If international cooperation is assumed, there is no reason why equally successful management systems cannot be developed for other areas.

Whereas in the past management-oriented research and collection of fishery statistics frequently have received much greater emphasis than biological studies, the latter are presently becoming of increasing importance in tuna management. For example, the study of species complexes, that is, groups of species that occur together and interact, are badly needed. In most of the world's tuna fisheries more than one species is taken during normal fishing operations, with an extreme being certain longline fisheries that often take as many as five species of tuna and tuna-like fishes in a single set of the gear. It is important for management to know what effect, if any, differential removal of one species has on the behavior and abundance of other species. For example, is it possible to maximize the yields of all species in a species complex simultaneously, or must some species be overexploited and others underexploited in order to best manage the complex as a whole? At present, quantitative aspects of relationships among species and the community dynamics of tunas are a vast unknown; to achieve insight into such relationships will require investments in scientific studies on a new order of magnitude.

In addition to relationships among tuna, studies of relationships between tuna and nontuna are also urgently needed. An important example is the tuna-porpoise relationship in the eastern Pacific. Tuna and porpoise occur together throughout the world's oceans, but only in the eastern Pacific do they occur in associated schools. This association, usually involving yellowfin, enables purse seiners to harvest the tuna more efficiently. By catching a school of porpoise, they increase their probability of catching a school of tuna. In fact, in some areas the fishermen are essentially capturing porpoise, but keeping only the tuna taken in association with the porpoise. They often succeed in releasing most or all of the porpoise alive, but unfortunately many are still inadvertently killed. This has created a serious and perplexing social problem. There is increasing political and legal pressure in the United States to totally curtail porpoise mortality for humane reasons. If it becomes necessary to eliminate porpoise fishing to achieve this objective and no effective alternative method is developed for harvesting tuna associated with porpoise, the eastern Pacific yellowfin catch could be reduced by as much as 60 percent or, based on the 1970–74 average catch, about 115,000 tons. At 1977 prices this would result in a loss of nearly 85 million dollars in dockside value or, very roughly, one-half billion dollars industry-wide. Because of the importance of the tuna-porpoise problem, it will be examined in detail in Chapter 12.

Distribution of the Catch among Harvesters

In most of the world tuna fisheries, production is limited by the supply of fish—that is, by a combination of factors that determines the capability of the oceans to produce tuna. In contrast to the limited supply of tuna, the fleets that harvest the resources have increased rapidly in recent years. Nations with developed fisheries have enlarged their fleets, while other nations are eager to enter the fishery. It is clear that the needs and expectations of all nations wishing to harvest tuna cannot possibly be fulfilled from the limited resources available. This results in sharp differences of opinion as to who should get what share of the available supply. The problem of how to distribute the catch among harvesting nations is probably the most important question in international tuna fisheries today, not only in the eastern Pacific where the management program is in jeopardy as a result of it, but also in other areas where it is acting as a serious impediment to implementation of management.

In considering the catch distribution problem, it is important to understand what motivates an individual or a nation to fish tuna. In the first place, tuna are not caught to feed directly to hungry people. Tuna are high on the ocean's food chain, and production of a ton of tuna requires between 10 and 100 tons of the animals that they eat. Also, it is expensive to harvest tuna. If feeding hungry people were the objective of tuna fishing, money used to catch them might be more efficiently spent to catch animals lower on the food chain. Instead, most tuna are caught because there exists a strong demand by people willing to pay a high price. Fishermen of nations with developed fleets fish for profit, not for food. They go after tuna because in the past it has provided an opportunity to make a better return on investment than any alternative undertaking. Developing nations also see in tuna fishing an opportunity for good returns to investment. In addition, to them it is a means for earning foreign exchange, creating jobs, and building industry.

At the crux of the catch distribution problem is a fundamental difference in political philosophy among nations interested in the tuna fishery which has already been touched upon. This fundamental difference lies between the "haves" and the "have-nots." The "haves" are nations that have developed major tuna fisheries, which, naturally enough, they do not want to curtail. The "have-nots" are nations without tuna fisheries (or with only small fisheries) which want and intend to develop large fisheries. By and large, the have-nots are developing coastal nations off whose shores tuna occur and are caught in substantial numbers.

The haves consist of six nations—Japan, the United States, the Republic of Korea, Spain, the Republic of China, and France—that take about 77 percent of the world catch of the principal market species of tuna. As a rough approximation, perhaps half of their catch is taken within 200 miles of the shorelines of have-not nations, with much of the remainder being taken on the high seas beyond 200 miles. Less than one-fifth is taken within their own coastal waters. In these six nations a world tuna industry has been created through great expenditures of time, effort, and money on development of technology for harvesting, processing, and marketing tuna products. The basic philosophy in these nations is generally that tuna, being highly migratory, should be the property of whoever first renders them to his use, and that their fleets should have ready access to fish wherever tuna occur.

The position of many of the haves is in direct opposition to that of many resource-adjacent have-nots who are determined to increase their shares of the world tuna catch at the expense of the haves. They feel that they should have preferential rights in harvesting the resources, and one suggestion has been that some portion of the catch should be distributed by allocating it to nations on the basis of resource adjacency. Even if national allocations could be agreed upon, however, significant problems would still have to be overcome. For example, tuna migrations and availability vary greatly from year to year depending on environmental conditions, and fleets of all participating nations would want to be able to operate wherever fish appeared in a particular year. Ways in which this might be worked out, assuming that nations extend their fisheries jurisdiction to 200 miles, will be discussed in detail in later sections.

For the eastern Pacific, where a detailed data base exists, the catch distribution problem can be explored in greater depth. It will be convenient to refer to nations adjacent to significant tuna producing waters as resource adjacent nations, or RANs. In the case of the eastern Pacific yellowfin and skipjack fishery, the RANs are, from north to south, Mexico, Guatemala, El Salvador, Nicaragua, Costa Rica, Panama, Colombia, Ecuador, and Peru. France is also classified as a RAN because of her possession, Clipperton Island. Other nations that participate in the eastern Pacific fishery will be referred to as non-RANs. In 1974 there were six such nations: the United States, Canada, Spain, Japan, Bermuda, and the Netherlands Antilles. Since 1974 Senegal, Venezuela, New Zealand, and the Congo have entered the fishery. Note that the classification of a nation as a RAN or a non-RAN depends on the specific fishery and area involved. For example, in the case of the northeast Pacific albacore fishery, the United States and Canada would be classified as RANs according to this scheme.

In 1974 six RANs had fleets of IATTC-monitored vessels that participated in the eastern Pacific fishery: Mexico, Costa Rica, Panama, Ecuador, Peru, and France. (Nicaragua began fishing in 1976, and France did not fish after 1975.) RANs took about 15 percent of the 1974 catch from the eastern Pacific, with the remainder being taken

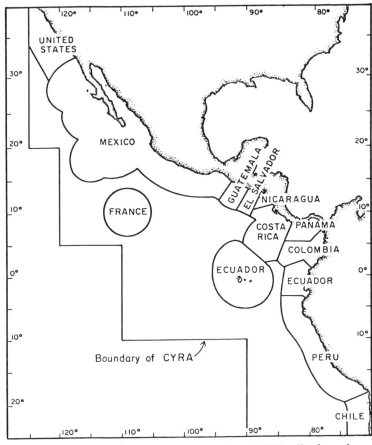

Figure 7. Approximate boundaries of areas extending 200 miles from shores of coastal nations in the eastern Pacific. Boundaries were determined by the IATTC staff following procedures described in the 1958 United Nations Convention on the Territorial Sea and Contiguous Zone. (See text for further clarifications)

by non-RANs, principally the United States. Of the total 1974 yellowfin and skipjack catch, 71 percent was taken within 200 miles of the RANs, and they see this 71 percent as a great source of potential wealth from the sea. To illustrate just how substantial this wealth really is, 200-mile zones were determined for each nation bordering on the CYRA (Fig. 7). The boundaries of these zones were constructed by the IATTC staff according to criteria set forth at the 1958 United Nations Convention on the Territorial Sea and Contiguous Zone in Geneva. It is to be emphasized that these 200-mile zones are defined for illustrative purposes only. In constructing them, the IATTC staff did not consult any of the nations involved, nor did it take cognizance of any agreements that might exist between nations on this matter. Hence the 200-mile zones indicated in Figure 7 should not be considered as precedent setting in any way. These reservations must be kept in mind in interpreting all of the illustrative examples based on Figure 7 throughout the remainder of this text.

Catches of yellowfin and skipjack tuna taken by the entire international fleet within the national 200-mile zones of Figure 7 are given for the years 1961–74 in Table 2. Catches beyond 200 miles but east of 155°W are also shown in Table 2. Three facts stand out. First, on the average in the years 1970–74, 45 percent of the combined yellowfin and skipjack catch in the eastern Pacific was taken within the coastal zones of Mexico, Ecuador, and Costa Rica, the three nations within whose zones the largest average catches were made. The remaining seven RANs had much smaller catches in their coastal zones amounting to only 19 percent of the total catch during the 1970–74 period. Second, since 1961, catches of tuna outside 200 miles have been increasing fairly steadily as a result of increased exploitation seaward. In the 1970–74 period these high seas catches averaged 36 percent of the total eastern Pacific catch. Third, catches within any single nation's coastal zone fluctuate widely from year to year. For example, yellowfin catches (in thousands of short tons) have varied between 40 and 3 off Costa Rica, between 14 and 2 off Peru, between 20 and 2 off Ecuador, between 61 and 15 off Mexico, between 29 and 0 off Panama, and between 7 and 0 off Nicaragua. Skipjack catches also have been highly variable.

If we consider these statistics as well as the fundamental philosophical differences among different categories of harvesting nations, it seems obvious that the problem of distributing the catch among users will be extremely difficult to resolve. The present system of access can hardly be sustained with nations unilaterally imposing and enforcing 200-mile limits. Nor does it appear likely, considering their dominant position in the market, that developed tuna fishing nations will passively agree to abandon all that they have developed over the years. Suffice it to say here: Some form of international cooperation appears

Yellowfin catch within 200 miles of --	1974	1973	1972	1971	1970	1969	1968	1967	1966	1965
Chile	0	0	0	0	0	0	0	0	166	0
Colombia	6,088	17,452	4,096	1,805	999	4,644	499	1,567	1,862	1,050
Costa Rica	39,554	23,587	18,033	21,712	10,662	9,480	10,229	4,009	8,989	8,750
Ecuador	18,500	15,143	16,125	19,456	12,239	7,241	11,843	9,888	19,898	8,146
El Salvador	14,228	1,539	469	1,455	276	501	10,316	1,539	3,733	4,123
France (Clipperton)	4,542	6,620	2,598	2,565	2,594	6,559	82	107	35	8
Guatemala	9,633	2,813	3,844	3,647	551	6,484	11,580	4,916	2,462	5,476
Mexico	46,262	22,314	27,316	22,740	61,424	40,248	42,774	43,328	14,701	34,092
Nicaragua	6,690	251	29	1,308	77	303	2,121	168	353	814
Panama	3,828	29,048	5,609	1,202	426	698	147	505	1,265	621
Peru	4,395	8,122	3,573	9,052	7,624	3,531	3,989	8,299	14,246	7,591
U.S.A.	9	0	0	25	9	0	10	0	0	0
Total within 200 miles	153,729	126,889	81,692	84,967	96,881	79,689	93,590	74,326	67,710	70,671
CYRA catch beyond 200 miles	35,701	48,992	70,505	28,189	45,821	46,823	21,023	15,323	23,447	19,372
Total CYRA catch	189,430	175,881	152,197	113,156	142,702	126,512	114,613	89,649	91,157	90,043
Catch outside CYRA and east of 150°W	41,686	48,804	44,763	23,370	29,775	19,177	0	0	0	0
Total eastern Pacific Ocean catch	231,116	224,685	196,960	136,526	172,477	145,689	114,613	89,649	91,157	90,043

Skipjack catch within 200 miles of --	1974	1973	1972	1971	1970	1969	1968	1967	1966	1965
Chile	0	0	0	0	0	0	0	0	32	85
Colombia	5,861	13,828	2,015	1,831	707	2,506	1,441	1,126	1,412	711
Costa Rica	28,943	3,337	299	18,554	968	995	14,823	204	1,153	1,953
Ecuador	12,383	8,424	11,218	42,382	15,264	29,601	25,852	58,606	36,568	53,735
El Salvador	4,775	0	3	1,011	0	5	7,252	152	13	184
France (Clipperton)	469	9	0	380	421	2,109	13	0	1	0
Guatemala	4,495	22	19	567	8	0	2,907	101	88	437
Mexico	9,241	8,799	9,942	16,647	26,417	8,628	7,747	37,422	2,915	7,428
Nicaragua	2,679	0	0	1,489	167	9	1,298	15	31	607
Panama	2,610	4,812	522	2,510	8	10	950	249	1,418	167
Peru	1,722	5,935	4,857	16,575	5,867	14,487	11,677	34,298	22,053	20,208
U.S.A.	0	0	0	0	37	5	3	11	295	24
Total within 200 miles	73,178	45,166	28,875	101,946	49,864	58,355	73,963	132,184	65,979	85,539
CYRA catch beyond 200 miles	10,554	2,277	6,910	11,295	5,552	5,837	3,721	305	547	583
Total CYRA catch	83,732	47,443	35,785	113,241	55,416	64,192	77,684	132,489	66,526	86,122
Catch outside CYRA and east of 150°W	2,848	1,436	1,248	1,163	6,418	981	0	0	0	0
Total eastern Pacific Ocean catch	86,580	48,879	37,033	114,404	61,834	65,173	77,684	132,489	66,526	86,122

Combined yellowfin & skipjack catch within 200 miles of--	1974	1973	1972	1971	1970	1969	1968	1967	1966	1965
Chile	0	0	0	0	0	0	0	0	198	85
Colombia	11,949	31,280	6,111	3,636	1,706	7,150	1,940	2,693	3,274	1,761
Costa Rica	68,497	26,924	18,332	40,266	11,630	10,475	25,052	4,213	10,142	10,703
Ecuador	30,883	23,567	27,343	61,838	27,503	36,842	37,695	68,494	56,466	61,881
El Salvador	19,003	1,539	472	2,466	276	506	17,568	1,691	3,746	4,307
France (Clipperton)	5,011	6,629	2,598	2,945	3,015	8,668	95	107	36	8
Guatemala	14,128	2,835	3,863	4,214	559	6,484	14,487	5,017	2,550	5,913
Mexico	55,503	31,113	37,258	39,387	87,841	48,876	50,521	80,750	17,616	41,520
Nicaragua	9,369	251	29	2,797	244	312	3,419	183	384	1,421
Panama	6,438	33,860	6,131	3,712	434	708	1,097	754	2,683	788
Peru	6,117	14,057	8,430	25,627	13,491	18,018	15,666	42,597	36,299	27,799
U.S.A.	9	0	0	25	46	5	13	11	295	24
Total within 200 miles	226,907	172,055	110,567	186,913	146,745	138,044	167,553	206,510	133,689	156,210
CYRA catch beyond 200 miles	46,255	51,269	77,415	39,484	51,373	52,660	24,744	15,628	23,994	19,955
Total CYRA catch	273,162	223,324	187,982	226,397	198,118	190,704	192,297	222,138	157,683	176,165
Catch outside CYRA and east of 150°W	44,534	50,240	46,011	24,533	36,193	20,158	0	0	0	0
Total eastern Pacific Ocean catch	317,696	273,564	233,993	250,930	234,311	210,862	192,297	222,138	157,683	176,165

NOTE: Sums of individual catches and proportions may not agree exactly with

200 MILES OF COASTAL NATIONS AND BEYOND 200 MILES IN THE EASTERN NATIONAL FLEET (Short tons)

1964	1963	1962	1961	Average catches			Proportions of total eastern Pacific catch		
				1970–1974	1966–1974	1961–1974	1970–1974	1966–1974	1961–1974
730	259	1,817	60	0	18	217	.0000	.0001	.0016
407	2,050	10,128	553	6,088	4,335	3,800	.0317	.0278	.0285
3,131	3,166	11,007	8,479	22,710	16,251	12,913	.1181	.1043	.0967
11,011	7,981	11,534	1,623	16,293	14,481	12,188	.0847	.0929	.0913
5,371	4,899	4,041	11,167	3,593	3,784	4,547	.0187	.0243	.0340
180	22	502	563	3,784	2,856	1,927	.0197	.0183	.0144
2,327	3,050	2,967	19,382	4,098	5,103	5,652	.0213	.0327	.0423
52,932	36,709	27,000	59,958	36,011	35,679	37,986	.1872	.2289	.2844
1,807	1,343	3,675	6,887	1,671	1,256	1,845	.0087	.0081	.0138
71	362	1,327	2,345	8,023	4,748	3,390	.0417	.0305	.0254
13,505	9,364	11,990	1,631	6,553	6,981	7,637	.0341	.0448	.0572
0	14	0	6	9	6	5	.0000	.0000	.0000
91,472	69,219	85,988	112,654	108,832	95,497	92,106	.5658	.6127	.6897
10,469	3,455	1,063	2,463	45,842	37,314	26,618	.2383	.2394	.1993
101,941	72,674	87,051	115,117	154,673	132,811	118,723	.8041	.8520	.8890
0	0	0	0	37,680	23,064	14,827	.1959	.1480	.1110
101,941	72,674	87,051	115,117	192,353	155,875	133,550	1.0000	1.0000	1.0000

1964	1963	1962	1961	Average catches			Proportions of total eastern Pacific catch		
				1970–1974	1966–1974	1961–1974	1970–1974	1966–1974	1961–1974
234	880	152	315	0	4	121	.0000	.0000	.0015
2,599	9,197	10,263	303	4,848	3,414	3,843	.0695	.0445	.0489
3,206	4,001	5,694	4,273	10,420	7,697	6,315	.1494	.1003	.0803
27,018	52,247	40,363	36,879	17,934	26,700	32,181	.2571	.3480	.4092
666	451	494	3,037	1,158	1,468	1,289	.0166	.0191	.0164
0	0	0	0	256	378	243	.0037	.0049	.0031
291	91	305	1,905	1,022	912	803	.0147	.0119	.0102
11,965	10,619	4,452	4,761	14,209	14,195	11,927	.2037	.1850	.1517
2,909	4,635	4,799	10,299	867	632	2,067	.0124	.0082	.0263
73	136	506	685	2,092	1,454	1,047	.0300	.0190	.0133
14,637	22,391	10,483	12,336	6,991	13,052	14,109	.1002	.1701	.1794
0	292	0	14	7	39	49	.0001	.0005	.0006
63,598	104,940	77,511	74,807	59,806	69,946	73,993	.8575	.9115	.9409
1,715	1,102	483	44	7,318	5,222	3,638	.1049	.0681	.0463
65,313	106,042	77,994	74,851	67,123	75,168	77,631	.9624	.9796	.9872
0	0	0	0	2,623	1,566	1,007	.0376	.0204	.0128
65,313	106,042	77,994	74,851	69,746	76,734	78,637	1.0000	1.0000	1.0000

1964	1963	1962	1961	Average catches			Proportions of total eastern Pacific catch		
				1970–1974	1966–1974	1961–1974	1970–1974	1966–1974	1961–1974
964	1,139	1,969	375	0	22	338	.0000	.0001	.0016
3,006	11,247	20,391	856	10,936	7,749	7,643	.0417	.0333	.0360
6,337	7,167	16,701	12,752	33,130	23,948	19,228	.1264	.1030	.0906
38,029	60,228	51,897	38,502	34,227	41,181	44,369	.1306	.1770	.2091
6,037	5,350	4,535	14,204	4,751	5,252	5,836	.0181	.0226	.0275
180	22	502	563	4,040	3,234	2,170	.0154	.0139	.0102
2,618	3,141	3,272	21,287	5,120	6,015	6,455	.0195	.0259	.0304
64,897	47,328	31,452	64,719	50,220	49,874	49,913	.1916	.2144	.2352
4,716	5,978	8,474	17,186	2,538	1,888	3,912	.0097	.0081	.0184
144	498	1,833	3,030	10,115	'6,202	4,436	.0386	.0267	.0209
28,142	31,755	22,473	13,967	13,544	20,034	21,746	.0517	.0861	.1025
0	306	0	20	16	45	54	.0001	.0002	.0003
155,070	174,159	163,499	187,461	168,637	165,443	166,099	.6434	.7112	.7828
12,184	4,557	1,546	2,507	53,159	42,536	30,255	.2028	.1829	.1426
167,254	178,716	165,045	189,968	221,797	207,978	196,354	.8462	.8941	.9254
0	0	0	0	40,302	24,630	15,834	.1538	.1059	.0746
167,254	178,716	165,045	189,968	262,099	232,608	212,187	1.0000	1.0000	1.0000

totals shown because of rounding.

to be necessary if the catch determination problem is to be satisfactorily resolved.

Economics and Fleet Carrying Capacity

A third major problem area confronting the world tuna industry is the rapid increase in the size of the fishing fleets. As fleets increase, competition for limited resources creates problems that could seriously jeopardize any arrangements that may be reached for management of the fishery and distribution of the catches.

It has been pointed out in discussing IATTC management in the eastern Pacific that the carrying capacity of the international fleet already far exceeds what is needed to harvest the available catch. As the international fleet grew from 44,000 tons of carrying capacity in the mid-1960s to 183,000 tons at the end of 1977, the annual catch per ton of carrying capacity dropped from 5.0 tons to less than 2.0 tons. In other words, vessels were limited to an average of less than two full trips per year. But at the present time a new 1,000-ton purse seiner must make about two to two and one-half trips per year just to break even. Thus, unless substantial new sources of tuna can be found and harvested (which seems unlikely over the short term), or unless prices paid for tuna rise very sharply, there will be vast economic disruption in the fishery including, with virtual certainty, many bankruptcies. Nothing illustrates the seriousness of the situation more clearly than the fact that since the early 1970s many major United States tuna-boat owners, foreseeing economic difficulties in the future, have sold their vessels to processors who seem to be integrating their industry vertically. Processors are apparently willing to purchase the vessels in order to insure that supplies of raw product will be available to their canneries.

This problem of over-capacity is also obvious on a worldwide basis. The carrying capacity of the entire world tuna fleet in 1971 was approximately 750,000 metric tons, while the world catch of the principal market species at that time was about 1.3 million metric tons. Thus, on a world basis, each ton of capacity produced about 1.7 tons of tuna. Since then fleets have continued to grow throughout the world.

It could be argued with some justification that a certain degree of overcapacity is beneficial because it forces excess gear to explore possibilities for exploitation of latent and underutilized resources.

Tuna vessels are designed specifically for catching tuna and tuna-like fish, however, and this limits possibilities considerably. There does not appear to be much potential for increased production of principal market species other than skipjack. But longline vessels do not harvest skipjack, and the surface fishing capacity now available could take skipjack catches much greater than those presently taken. Billfish resources are of limited size and already fully exploited. Some possibilities may exist, however, for increasing production of secondary market species.

In addition to not making good economic sense, continued unrestricted fleet growth also creates serious political problems by exacerbating the situation regarding distribution of the catch among users. As fleets increase beyond the capability of the tuna stocks to fill their holds, disputes over who should get what share of the limited available harvest could intensify to the point where they become so dominant in everyone's mind that finding solutions to other important problems becomes impossible. To prevent this from happening, there is a strong need to limit the number of tuna vessels being built. Though most agree that such a need exists, it will be extremely difficult to control fleet size because of conflicting interests among nations. Those with large fleets want any controls on future vessel construction to apply to all nations equally. This, in effect, would insure their continued dominance in the fishery. Nations without large fleets, especially the RANs, believe that controls should initially apply only to nations with developed fleets. This would allow them an opportunity to develop their own industries. These conflicting viewpoints among nations regarding fleet limitation are closely related to their conflicting attitudes on catch allocation.

Whatever is done over the short term, limitations must ultimately apply equally to all participants if effective control of fleet growth is to be achieved. Otherwise the result will be failure. The history of the Japanese longline fishery provides a vivid example. From about 1950 through 1968 Japan took essentially 100 percent of all tuna captured with longline gear. In a cooperative industry-government program, the number of vessels entering the fishery was strictly controlled. Controls were based on conditions in the fishery and were designed to maintain abundant stocks of tuna and resultant high fishing success. In this way vessels were virtually guaranteed catches that would result in a return on investment exceeding a minimum established level. In the late 1960s the Republics of China and Korea, frequently using capital from Japanese sources, began to purchase longline vessels. Competing directly with the Japanese flag fleet, they increased their fleets as rapidly as possible in order to maximize their shares of the catch. This resulted in excessive fishing effort. Tuna stocks available to longline gear declined, and hooking rates, which measure fishing success, fell off sharply for all participants including Japan. As a result, since 1975 substantial segments of the international longline fleet have been tied up in port because catches are not sufficient to defray expenses, which have increased, especially in regards to fuel costs. Yet, ironically, longline vessels still continue to be built.

The great mobility of the world's tuna fleets complicates the problem of developing controls on fleet growth. For example, after termination of unrestricted fishing for yellowfin in the eastern Pacific, many vessels customarily transfer to the Atlantic. Similarly, Atlantic vessels sometimes fish in the eastern Pacific during the first half of the year. It seems reasonable that if the eastern Pacific fleet is to be set at a certain tonnage, then when this tonnage is determined account should be taken of the extent to which vessels in the fleet will also fish in the Atlantic. The basis for doing so, however, is not clear-cut.

Another difficulty that must be considered in limiting fleet growth is how to deal with vessels wishing to enter a fishery for the first time. This can be called the new-entrant problem. With respect to a specific fishery, new-entrant vessels can be classified as being from either a RAN or a non-RAN.

The non-RAN vessel category could be subdivided into (1) new-entrant vessels from non-RANs whose fleets have historically participated in the fishery in question; and (2) new-entrant vessels from non-RANs whose fleets are entering the fishery for the first time. If the number of new-entrant vessels must be restricted it could be argued that vessels from historically participating non-RANs should be favored over vessels from non-RANs that have not previously participated. However, from a worldwide perspective, this would probably be ill advised. While a non-RAN might benefit from being a historical participant in one fishery, the precedent established could be damaging when the nation wished to enter a fishery elsewhere for the first time. Most nations would probably prefer to maintain the greatest possible degree of flexibility with regard to entering new fisheries.

On the assumption that shares of the allowable catch have been allocated to RANs, two categories can be distinguished. First, there are

RANs whose fleets are undeveloped or underdeveloped in the sense that they are not capable of fully harvesting their allocations. New-entrant vessels joining the fleets of these RANs should probably be favored over all other vessels if the number of new entrants must be limited. The second category would include all RANs whose fleets are developed to the point where they can fully harvest their allocations. New-entrant vessels from these RANs should probably be treated the same as new-entrant non-RAN vessels and not receive any special preference.

Enforcement of Conservation Regulations

Assume that the problem areas already considered can be somehow resolved, and that regulations for conservation of tuna resources have been adopted. If they are to be effective, they must be adequately enforced, and enforcement must be equally stringent with respect to vessels of all nations participating in the fishery. Through 1977, tuna fishing regulations in the eastern Pacific have been limited to an overall catch quota for yellowfin. In the Atlantic there have been size limits for yellowfin and northern bluefin, and certain nations have cooperated to limit fishing mortality on northern bluefin. Each individual nation participating in these fisheries has been responsible for enforcing conservation regulations with respect to its own vessels, but, unfortunately, some nations have not been able to do so effectively. Naturally this prompts nations that do enforce regulations to raise the question: Why should our vessels be restricted when vessels of other nations are not similarly restricted? To resolve this question, it is clear that some sort of international enforcement system must be established. This will be far from easy to do, for enforcement gets into the very sensitive area of national sovereignty. However, one optimistic point can be made: If an effective solution can be found for one fishery, it is likely that a similar solution can apply to other fisheries.

The form of an enforcement scheme will depend to a large extent on the nature of the conservation regulations in effect. For example, in the Atlantic, where there are minimum size limits, an adequate enforcement system would require a workable catch inspection system. At present, such a system does not exist, and there is widespread concern that minimum size limits are not being observed.

The situation in the eastern Pacific is more complicated and will be examined in some detail. As noted in Chapter 3, the yellowfin catch quota applies to a large sub-area of the eastern Pacific fishery known as the Commission's Yellowfin Regulatory Area (CYRA). Fishing for yellowfin tuna in the CYRA begins on January 1 of each year, and as the season progresses the IATTC staff monitors the catches of the fleet at sea on a daily basis via radio reporting. When the cumulative catch of yellowfin plus the expected catch in the CYRA after closure equals the quota, the season closes. Vessels fishing in the CYRA after closure must be completing unregulated trips, fishing under some special allowance, or fishing for species other than yellowfin. Vessels in the latter category are allowed an incidental catch of yellowfin of up to 15 percent of their total catch. Vessels fishing west of the CYRA or in the Atlantic are not restricted in their yellowfin catches.

To enforce the conservation regulations effectively, surveillance of all vessels fishing in the eastern Pacific outside the CYRA after closure is necessary. Otherwise a vessel could report that it was fishing outside the CYRA when in reality it was fishing illegally for yellowfin inside the area. There must also be a capability to inspect catches of vessels returning to port from regulated trips in the CYRA (i.e., fishing for species other than yellowfin) to insure that illegally caught yellowfin are not being sold as skipjack, bluefin, bigeye, or some other species for which catch is not controlled. Unfortunately, violations of both kinds do occur because the capability to enforce conservation regulations varies from nation to nation. Some technologically advanced nations are able to routinely maintain surveillance over their vessels at sea using advanced electronic equipment while other less advantaged nations have no such capability. Any nation has the capability to inspect the catches of its own flag vessels if they unload in their own territory, but inspection of unloadings made in other nations is impossible without state permission. Such permission is often uncertain, being dependent upon political relations among nations. Thus, some nations with a sincere desire to effectively enforce conservation regulations to which they have agreed may not be able to do so. This situation jeopardizes the entire conservation program in the eastern Pacific.

In an attempt to resolve the problem of unequal enforcement capability, the United States has proposed a reciprocal port-of-landing inspection scheme and has offered to routinely determine positions of foreign flag vessels by radio triangulation. Any violations would be reported to the flag nations involved, who would retain responsibility for taking appropriate legal action. But other nations have declined to enter into such arrangements because they believe to do so would impinge upon their sovereign rights.

An alternative approach, and one that could be applied either in the eastern Pacific or on a worldwide basis, would be for individual nations to delegate the responsibility for surveillance and catch inspection to an international body representing all states. This body could be established at any suitable location in the world, and would have an internationally recruited staff of inspectors, technicians, and other personnel. Its basic responsibilities would be to determine positions of tuna vessels while they were at sea and to inspect catches of unloading vessels. If violations were detected, they would be reported to the nation whose flag the violating vessel was flying. Administration of legal action would remain the responsibility of the sovereign states. To insure that punitive actions imposed by participating nations are equitable, guidelines for fines and other penalties could be established by member nations through the international body.

Determining positions of vessels at sea could be accomplished easily and effectively through satellite technology. A small radio transponder could be placed aboard each vessel at a modest cost. This "black box" transponder would be interrogated frequently using existing satellite technology. The resulting data would be relayed to a shore base where each vessel's position could be accurately fixed. It might also be possible to monitor a vessel's gyrocompass and main drive shaft via the black box to determine its course and speed. A small computer system for interrogating the satellite, determining vessel positions, courses, and speeds, and routinely printing out this information could be acquired, and only a minimal staff would be required to operate the system. The black boxes would be transistorized and almost failure proof. They would normally operate on ship's power, but could automatically shift to a back-up battery system so that power outages would not affect transmission.

Catches of unloading vessels would be examined by international inspectors to determine the species and amounts caught. They would also check for violations of any size regulations that might be in effect. To reduce the number of inspectors needed, a scheme could be developed for selecting vessels to be inspected on a random basis. Also, inspectors could move from port to port as conditions dictated.

The cost of operating an international body with responsibility for surveillance and catch inspection would be rather nominal. The body would not have to concern itself with apprehension of violators and thus would not need to maintain an expensive fleet of patrol vessels. Of course, nothing could be done to prevent vessels of noncooperating nations from fishing beyond the juridical zones of cooperating nations, but such vessels would be excluded from coastal waters of cooperating nations. This by itself would provide a strong incentive for cooperation. Preventing infringements of the coastal zone by vessels of noncooperating nations would remain the responsibility of cooperating coastal nations.

It would probably be most efficient for an international surveillance and catch inspection body to be established as a branch or affiliate of an international agency with overall responsibility for conservation and management of tuna. It should be emphasized that all information gathered for the purposes of research and management by such an international agency must be kept absolutely separate from its enforcement-related functions. If the same individuals who are responsible for collecting research data also collect enforcement information, or if data collected for research purposes are ever used for enforcement purposes, then the flow of data from harvesters and processors to the scientists will be adversely affected. Under careful controls, however, certain information gathered by an enforcement branch might be made available for research studies.

The importance of separating international enforcement activities from research activities is well exemplified by the experience of the IATTC in the eastern Pacific. A good share of all the data used in IATTC research (e.g., catch weights, catch locations and dates, gear employed, species caught, size frequency samples, unloading data, processor sales receipts, etc.) are obtained through cooperation with industry. The data are provided with the understanding that they will be used only for scientific studies, and that they will not be made available to anyone outside the scientific field in any form that could reveal the activities of an individual vessel, vessel broker, processor, or wholesaler. In fact, this concept was written into the convention of the IATTC along with its rules of procedure. Further protection is provided by the fact that IATTC records and data are inviolable because it is an international organization. Because the staff of the commission, at the direction of the Commissioners, has always scrupulously observed this confidentiality policy, it has been possible to establish one of the most comprehensive fishery data bases in the world. This data base has been of fundamental importance in maintaining the research and management program. Mixing research with enforcement could seriously jeopardize the opportunity to collect essential data and any research use of enforcement data should be undertaken with extreme caution.

From what has been said, it is clear that the capability exists to solve the enforcement problem. Only the political will to do so is required. In fact, solution of the enforcement problem should be relatively straightforward if some internationally acceptable solution to the catch distribution problem can be achieved.

5   Conservation and Tuna Management Philosophy

The nature of the world's tuna fisheries and their present management have been described. Also the critical problems plaguing nations that fish for tuna have been identified. Before we consider possible solutions to these problems, it is appropriate to pause briefly to consider just what we mean when we talk about conserving tuna resources.

The term "conservation" has been often defined and is much used, being especially in vogue these days. Yet its meaning remains unclear. It has something to do with natural resources and how they are used or not used, how they are exploited or preserved. However, almost everyone would agree that conservation involves the wise and rational use of limited natural resources. The difficulty is in deciding what constitutes "wise and rational use." At one extreme are dedicated preservationists who seek to secure significant portions of natural resources in a virgin, undisturbed state; they would conserve aesthetic qualities, wildlife habitat, wilderness recreation values, opportunities for scientific studies, and other noneconomic uses. At the other extreme are those who plan and manage the extraction of wealth from natural resources; for them, wise use means providing society with needed goods and raw materials, at the same time maximizing economic gains.

To narrow attention to the fisheries field, an important definition of conservation was developed by the 1958 Geneva Convention on the Law of the Sea (Article 2 of the Convention on Fishing and Conservation of the Living Resources of the High Seas): ". . . the expression 'conservation of the living resources of the high seas' means the aggregate of the measures rendering possible the optimum sustainable yield from those resources so as to secure a maximum supply of food and other marine products." The emphasis in this definition is on exploitation of marine resources, and while such a definition might

40

not be adopted in today's circumstances (consider porpoise resources for example), it certainly is in keeping with the dominant thrust of fisheries management philosophy over the last thirty years or so.

Fisheries conservation can operate at various levels. One possible level is absolute preservation, in which the resource being conserved is left unexploited in its virgin state for aesthetic or other reasons. At the opposite level is exploitation so intense that the resource is reduced to a very low level, with production sustained at only a fraction of the potential production. In dealing with renewable resources such as fisheries, society must choose the level of conservation it wishes to pursue during any period of time. But in making the choice there is a minimum level of conservation below which no generation of users has the right to go. At this minimum level, options for future use must be retained. No generation has the right to exploit renewable resources in a way that excludes the possibility of their use by succeeding generations.

In the case of tuna, the minimum level of conservation, the most fundamental management requirement, is to maintain reproductively viable stocks. Stocks must not be driven to levels where extinction is threatened or which preclude recovery to former abundance in a reasonable period of time. Within this limitation the level of conservation that society chooses is determined by current needs, be they wisely or unwisely perceived. During recent decades, and in accord with the Geneva definition of conservation noted above, a commonly held opinion has been that tuna stocks should be maintained at levels from which maximum physical yields can be taken on a sustained basis. Indeed, this is the stated management objective in the convention establishing the IATTC. There has been considerable controversy, however, as to whether or not the maximum sustainable yield philosophy provides an appropriate basis for establishing management goals, and there has been a tendency by many to substitute the concept of optimum yield. The distinction is an important one: The term *optimum,* though vague, clearly encompasses objectives (such as those based on social and aesthetic considerations) that go beyond purely economic and biological objectives. Nevertheless, maximum sustainable yield is an important element in the set of possible optimum yields. In examples given in the following chapters, maximum sustainable yield is assumed to be the objective of tuna management.

The preparation of this study has been motivated by a basic underlying assumption: that the nations of the world seriously consider conservation of tuna stocks to be a desirable objective, and that they are willing to cooperate in implementing some form of management to attain this objective on a lasting basis. In view of the philosophical conflicts between RANs and non-RANs, it is clear that cooperation to achieve a meaningful conservation program implies sacrifice and compromise by all with an interest in the resource to be conserved. In the remainder of this book, alternative tuna management systems will be explored with these points in mind.

In evaluating management alternatives, the focus will be on determining whether or not they provide adequate solutions to the basic tuna management problems outlined earlier, especially the catch distribution problem. The following approaches to management will be considered and evaluated: (1) control by RANs of all fishing within their individual 200-mile coastal zones with no international agreement for distributing the catch; (2) extension and modification of the overall quota system presently in effect in the eastern Pacific; (3) partial allocation of the catch to RANs based on resource adjacency with open access and a system of participant fees; (4) control of fishing by regional coalitions of various kinds; (5) allocation of the entire catch among both RANs and non-RANs; and (6) catch allocation through international competitive bidding. These systems will be discussed primarily from the point of view of their possible implementation in the eastern Pacific, and specific examples will be based on experience there, but the general principles developed should apply equally to other tuna fisheries of the world.

6   Control to 200 Miles
       by Individual Coastal States

As has been noted, zones of economic influence throughout the world are being extended seaward to 200 miles. The question of who should have jurisdiction over highly migratory pelagic species such as tuna as they pass through the economic zones of various countries has been the subject of much debate at the third United Nations Conference on the Law of the Sea and elsewhere. Although the question is still unresolved, it is possible that jurisdiction will be in the hands of individual RANs. The consequences of individual RAN control will be explored in this chapter. A related possibility, that RANs might enter into some type of regional coalition, will be discussed in Chapter 9.

It is reasonable to assume that RANs will want to maximize catches of tuna in their own coastal zones, either by harvesting the resource themselves or by making arrangements for foreign flag vessels to do so. If each RAN independently determined a desired catch level for its own coastal zone, the total desired catch for the coastal zones of all nations combined would be impossible to estimate. It is safe to assume, however, that the figure would not be less than that which would be arrived at if each RAN hoped, either by itself or by licensing others, to achieve a catch equaling the recent average catch taken in its own coastal zone by the entire international fleet. Applying this conservative assumption to the eastern Pacific, data for the 1970–74 period in Table 2 show the following: For yellowfin the total desired RAN catch would be 109,000 tons (70 percent of the CYRA catch, 57 percent of the eastern Pacific catch); for skipjack, 60,000 tons (89 percent of the CYRA catch, 86 percent of the eastern Pacific catch); and for the two species combined, 160,000 tons (76 percent of the CYRA catch, 64 percent of the eastern Pacific catch). Comparable figures for the 1975–76 period are as follows: for yellowfin, 127,000 tons (67 percent of the CYRA catch, 53 percent of the eastern Pacific catch); for skipjack, 104,000 tons (76 percent of the CYRA catch, 75 percent of the eastern Pacific catch). Whatever base years are used, these are substantial expectations, and there are a number of approaches that the RANs could adopt in trying to fulfill them while at the same time maintaining full control in their individual coastal zones.

Exclusive National Fishing Zones

It is logical and instructive to consider first the most extreme form of access restriction by assuming that each RAN excludes fleets of all other nations from its coastal zone with the intention of developing its own fleet. If this were actually done, non-RAN fleets which in 1975 took nearly 75 percent of the total yellowfin and skipjack catch would be excluded from fishing grounds which during the 1970–74 period yielded 64 percent of the total catch. These figures imply a sharp reduction in non-RAN catches, and it seems obvious and inevitable that non-RAN fleets would react by maximizing their outside catches. Also it seems highly unlikely that they would be willing to cooperate in any conservation program that involved limiting their fishing beyond 200 miles.

How would an intensely competitive international fishery operating without restriction beyond 200 miles affect yellowfin stocks? Yellowfin are migratory throughout the eastern Pacific and their pattern of distribution varies seasonally and from year to year. This is reflected by the fact that yellowfin catches within 200 miles have varied from 68,000 to 154,000 tons during the decade from 1965 through 1974 while catches beyond 200 miles have varied from 15,000 to 115,000 tons. Tagging data also indicate significant onshore-offshore mixing. Thus, it appears certain that intensified fishing could increase yellowfin catches beyond 200 miles. It is likely that such an intense fishery would be fairly competitive with coastal zone fisheries, and that potential production of yellowfin from within the coastal zone could be reduced, even in the unlikely event that conservation measures were established by individual RANs.

The biological consequences of an intense outside fishery are difficult to predict, but the ultimate result would quite likely be some degree of overexploitation of yellowfin. Just how far yields might fall below their potential maximum level depends on the spawner-recruit (S-R) relationship for the stock and the yellowfin mixing rates between inshore and offshore areas. As compared to the inshore fishery, the offshore fishery tends to take larger fish, and if there is a strong S-R relationship, sharp reduction of outside spawning stocks would result in lower recruitment and reduced catches. Ultimately this would mean reduced catches for all. The exact relationship between stock size and subsequent recruitment is not known over the entire range of stock sizes, but below some level it seems nearly certain that recruitment will begin to fall, perhaps very rapidly. Under these circumstances it is obviously prudent to assume a relatively strong S-R relationship for yellowfin.

The situation with regard to skipjack differs in important respects from the yellowfin situation. A much greater percentage of the eastern Pacific skipjack catch is taken within 200 miles (86 percent for skipjack versus 57 percent for yellowfin for the 1970–74 period). This means, at least initially, that an offshore international fleet of non-RAN vessels would be virtually excluded from the skipjack fishery. However, skipjack enter the coastal waters of the eastern Pacific seasonally from the central Pacific. It is possible that if excluded from traditional skipjack grounds, non-RAN fleets might develop techniques to successfully intercept and harvest skipjack before they entered the coastal zone, even though costs might be greater. (As already noted, a tendency to take a greater proportion of the skipjack catch beyond two hundred miles has been developing since 1974.) This could substantially reduce the abundance of fish inshore. The ultimate effect of such potential competition for skipjack is uncertain, especially since relationships to skipjack stocks in other parts of the Pacific are not entirely clear. Overexploitation seems less of a threat, however, than in the case of yellowfin.

In addition to competition from non-RAN fleets fishing beyond 200 miles, all RANs face serious problems if they exclude one another from their respective coastal zones. Table 2 shows that annual catches from the coastal zones of some nations have averaged less than 5,000 tons even though the international fleet that took these catches was extremely large. A small fleet fishing exclusively in the coastal zone of one of these nations would be hard pressed to capture 2,000–3,000 tons annually,hardly enough to keep even a single large vessel operating profitably. Obviously such nations could not develop major tuna fisheries. Nations with larger average catches in their coastal zones would also face great difficulties because catches of both yellowfin and skipjack are highly variable. For a country like Mexico which is building a large fleet, that fleet in most years would have to fish in the coastal zones of other nations. Costa Rica, Ecuador, Nicaragua, Peru, and Panama are also developing modern fleets of large vessels that could not possibly make adequate catches every year if restricted to their own coastal zones.

The Ecuadorian skipjack fishery vividly illustrates the serious problems that RANs will encounter if they build large fleets to take tuna in quantities based on local catches in good years but are unable to move these fleets into waters under other jurisdictions when local fishing is poor. The latitude at which skipjack enter coastal waters in the eastern Pacific has varied from year to year. In certain years, such as 1961–63, 1965–67, and 1971, they have entered and concentrated in waters off Ecuador and northern Peru, and catches there by the international fleet have ranged from just under 50,000 tons to over 90,000 tons. In other years, such as 1970 and 1972–74, they have entered to the north off Central America, and catches in the southern area have dropped sharply, often falling below 20,000 tons. During the 1960s, when large concentrations of skipjack were seasonally available close to Ecuadorian ports, a large fleet of small vessels was developed. Canneries were built and support industries grew as landings rose to more than 20,000 tons. In 1972 and 1973, when the seasonal skipjack runs shifted beyond the range of the Ecuadorian fleet of small vessels, landings dropped to below 6,000 tons and economic hardship beset much of the industry. Low production and poor earnings forced many vessels and some shore facilities to curtail operations. Although this economic hardship resulted from the incapability of the small vessels to fish distant waters, jurisdictional barriers would result in similar problems for larger vessels. Substantial new investment in larger, more far-ranging vessels will be required to increase and sustain Ecuadorian production, and these vessels will have to have access to all areas where tuna concentrate.

With the preceeding considerations in mind, what changes would be likely in fleet composition and catches under the exclusive national fishery approach? Initially, the total yellowfin and skipjack catch would almost certainly drop because major portions of the fishery would be closed to the bulk of the fleet. As has been noted, however, the catch taken beyond 200 miles would go up because much greater effort would be concentrated there than at present. Offshore catches might continue to increase somewhat, at least for a while, as offshore fishing techniques develop and improve, but it is unlikely that increases in the offshore catch could ever offset losses in inshore catches which would be down sharply because of exclusion of non-RAN fleets. If RAN fleets increased in size, their inshore catches would go up, but it is unlikely they would ever reach the level of present inshore catches because these are taken by the entire international fleet. Overall the combined inshore and offshore yellowfin catch could conceivably approach the level it has been at in recent years. The total catch would go down substantially, however, unless offshore skipjack catches could be sharply increased. With no management, the great offshore effort on yellowfin could result in overfishing depending upon the S-R relationship. Possible consequences of excluding non-RAN fleets from the 200-mile zone are explored more quantitatively in Chapter 9 in which it is assumed that RANs form a coalition among themselves to provide for mutual access to one another's waters.

The manner in which RAN fleets grow would make a difference. If growth were accomplished by flag changes involving transfer of non-RAN vessels to RAN fleets, then these fleets could grow rapidly. Competition for the unmanaged yellowfin resource would be extreme, with non-RAN fleets concentrating on larger fish offshore while RAN fleets concentrated on the primarily younger fish inshore. If RANs must develop their fleets by means of new construction, changes would be more gradual. In fact, it would not be surprising if investment capital proved hard to find because of such serious problems as lack of access to foreign coastal zones, competition from outside fleets, and possible overexploitation. Problems for RAN fleets would be accentuated when inside fishing was poor, especially in years when one or both species of tuna appeared in below-average abundance. In such cases RAN fleets might fish outside the 200-mile zone, competing with non-RAN fleets there.

In conclusion, creation of exclusive fishing zones can hardly be considered an acceptable approach to tuna management. It fails to satisfactorily resolve the catch distribution problem, and it makes management of the resources impossible. With no management, damage to the resources as a result of overfishing is quite possible and economic hardship virtually inevitable for all concerned. With drawbacks of this magnitude, the problems of enforcement that would face the RANs need not even be discussed.

Licensing by Individual Nations

In seeking a means for deriving benefit from tuna resources occurring in their coastal zones, RANs will undoubtedly consider licensing systems as an alternative to maintaining exclusive fishing zones. In fact, extension of jurisdiction to 200 miles and establishment of licensing systems for foreign flag vessels is already an accomplished fact in some Latin American countries.

The motives for licensing foreign flag vessels are fairly obvious. Licensing provides a means for RANs without their own fleets to derive benefits from tuna resources in their coastal zones. Similarly, nations with modest fleets can benefit from both catches and license fee revenues in years of abundance. More important, if all RANs institute licensing systems, then the developing fleets of each RAN can fish tuna wherever they become abundant within the 200-mile zone in a particular year, provided, of course, that they can afford the license fees. Thus licensing would appear to offer a possible solution to the problem of maintaining access to a resource that is highly variable in its distribution both within a given year and also from year to year.

A licensing system could be implemented in different ways, depending on the objectives of a particular RAN. A nation without its own fleet would want to maximize revenues from foreign flag vessels, while a nation developing a fleet of its own would probably structure a licensing system to favor its own vessels if possible. Favoritism could take several forms. For example, it is unlikely that a nation would require its own flag vessels to purchase licenses, and this would obviously give them an economic advantage over foreign flag vessels that would pay. Another possibility that might appeal to some nations would be to license only vessels of other RANs, thus excluding the non-RAN fleet. Assuming that all RANs allowed each other's vessels to purchase licenses, this strategy would preserve access for themselves throughout the 200-mile zone while significantly limiting competition from the non-RAN fleet. However, this strategy would not suit RANs that desire to maximize catches and revenues from within their coastal zones. They logically would want to encourage the non-RAN fleet to buy licenses. Another conceivable strategy to assure a sufficient supply of fish for one's own fleet would be to limit either the number of licenses sold or the catch by licensed vessels. However, if the time lags involved in deciding how many licenses or how much catch are balanced against the rapidity with which tuna availability can change, establishment of such limits seems impractical.

Regardless of how various national licensing schemes are set up and implemented, major problems can be anticipated. They fall into three general categories: (1) problems associated with determining the basis for license fees and their amount; (2) problems related to enforcement of a licensing system; and (3) problems stemming from the lack of an effective system for management of the resource as a whole.

Fees in existing licensing systems are generally based on a vessel's registered net tonnage. Thus, a vessel must pay for the privelege to fish whether it catches anything or not. This clearly adds to the risk of fishing, because license fees increase fishing costs substantially with no assurance of a commensurate increase in production. In a single season, indeed even on a single trip, a modern purse seiner frequently fishes off the shores of several nations. If a license must be purchased from each nation in order to fish, and if each license costs as much as an Ecuadorian or Peruvian license cost during 1977, it is hard to imagine how licensing can be successful. For example, an Ecuadorian license for a single trip during 1977 cost a seiner with a carrying capacity of 1,000 tons approximately thirty thousand dollars, and Peruvian license cost fifty thousand dollars. The years 1973 and 1974 provide a pertinent case in point. In these two years fishing was concentrated in inshore waters off Central America from Panama to Guatemala. In order to follow the fish and make a full load, vessels during a single trip often fished in the coastal waters of five different nations. If all these nations had required licenses and had charged for them at Ecuadorian or Peruvian levels, it would have been economically impossible for a vessel to pay such fees. It is important to note that this problem would have applied to RAN fleets as well as non-RAN fleets. Of the nations fishing in this Central American area in 1973 and 1974, only two were Central American nations (Panama and Costa Rica), while four of the remaining nations (Mexico, Ecuador, Peru, and France) were RANs that would have had to buy licenses. In fact, even the Panamanian and Costa Rican vessels would have faced a licensing problem.

If a vessel could not afford to pay for, say, half a dozen licenses during a year, then it would be left with the choice of buying no license at all and fishing beyond 200 miles, or else buying licenses to fish in the coastal waters of just one or two nations. A few RANs blessed with large resources such as Mexico, Ecuador, and Costa Rica might sell a fair number of licenses, but other nations would sell few licenses or none at all.

As an alternative to license fees based on registered net tonnage, a coastal state could license boats to fish for a nominal fee or even for no fee at all, and instead collect a catch tax or participant fee that would be based upon actual catches. A participant fee system would mean that costs and catches would increase together. Because this would reduce risks, fleet movement and the distribution of fishing effort would be less inhibited, quite possibly resulting ultimately in greater revenues to coastal states. The participant fee concept is an important one that is fully discussed in Chapter 8.

The next problem area to consider is enforcement. To be effective, any licensing system would have to be accompanied by an adequate enforcement system. Such a system would require both a surveillance capability to detect unlicensed vessels fishing in one's coastal zone and a capability to apprehend and detain violators when they are detected. Surveillance of a vast sea area generally relies on a combination of aerial and surface reconnaissance. Apprehension and detention require that a fleet of patrol vessels be maintained. Because these are all costly activities, the cost of developing and operating an enforcement system which would have a high probability of detecting and apprehending a violator could easily exceed license revenues. Hence, for effectiveness, enforcement would probably depend instead upon imposition of extremely severe penalties when violators are caught. Indeed, both in the United States and in Latin America, increasingly severe penalties have been imposed for foreign fishing violations in waters over which jurisdiction is claimed.

Enforcement would be especially difficult if a nation decided to set up a participant fee system instead of licensing the right to fish. A participant fee system would require quantitative monitoring of catches, and it is obviously more difficult and costly to keep track of exact amounts caught than simply to note the presence of a fishing vessel in one's coastal zone.

Even if RANs could develop licensing systems that would be enforceable and produce adequate income, one serious problem would remain. With each nation independently establishing its own scheme for licensing and with the fishery being open to all beyond 200 miles, it seems unlikely that a viable resource management program could be maintained. In order to see why management is likely to fail under a licensing approach, it is necessary to consider the level of license fees, because distribution of fishing effort would depend strongly on this factor.

If license fees were so low as to not unduly inhibit vessel movement in the coastal zone, then concentration of fishing effort beyond 200 miles would probably be no greater than at present, and fleets could follow the fish more or less as they do now. Under these conditions, with the whole resource relatively accessible to all harvesters, the large and still growing international fleet would quite likely overfish at least the yellowfin stocks if no international conservation agreement (such as the existing IATTC agreement) could be maintained. On the other hand, it is possible that a management agreement might be maintained under this low-fees licensing scenario, in which case overfishing could be avoided. However, the likelihood of low fees being established seems almost nil. In the first place RANs would receive only limited revenue from fees set so low as to not inhibit fleet movement, but enforcement and administration costs would be as great as ever. Additionally, a fully mobile international fleet with access to the entire available resource could pose a real competitive threat to RANs interested in developing their own fleets. For these reasons alone establishment of a low-fees licensing system seems un- likely, but there is a further important reason. By imposing a licensing system, a RAN could establish its authority to regulate tuna fishing in its coastal zone. If fees were very low, however, foreign flag vessels would gain access to coastal tuna resources without any formal recog- nition being given to the RAN belief that they are entitled to special shares of the catch based on their resource adjacency. Thus, setting low license fees would amount to a *de facto* abandonnent of a strongly held philosophy concerning coastal states' rights to utilize marine resources occurring in their waters.

Since low license fees do not appear likely, the prospects for main- taining an adequate management system must be considered under the assumption that all nations would set license fees at levels compa- rable to those charged by some nations now. At these levels, as has been pointed out, it would be economically impossible to purchase more than one or two licenses per trip, and many vessels in both RAN and non-RAN fleets might not buy any licenses at all. The resulting limitations on fleet movement would lead to an overall situation very similar to that described in considering the exclusive national fishing zones approach. Vessels fishing beyond 200 miles would not be willing to limit their catches if they were largely excluded from inshore waters by high license fees. Also, RANs would want to maximize catches in their coastal waters whenever fish became available. Under these circumstances there could be no effective management, and one can roughly anticipate what would result. Beyond 200 miles the yellowfin catch would initially increase as non-RAN fleets and fleets of resource-poor RANs competed on an unrestricted basis. Inside the coastal zones catches of both species would decline, with the amount of decline depending on just how high (and therefore, how exclusionary) the license fees were. Later the trend of inside catches would depend on the rapidity of RAN fleet growth, trends in license sales, and possible overexploitation of yellowfin by the intense outside fishery.

In summary, while effective management is theoretically possible if license fees are very low and do not impede fleet movement, such low fees would be unacceptable to RANs for both economic and philosophical reasons. On the other hand, high licensing fees would inhibit fleet movement, encourage uncontrolled fishing beyond 200 miles, and make management impossible. Declining catches, possible overexploitation of yellowfin, and economic chaos for all would result under these circumstances. Thus in terms of solving the catch distribution problem, licensing by individual nations seems just as unpromising as maintaining entirely exclusive national fishing zones.

7   Extension of the Present
Eastern Pacific
Overall Quota System

Because management systems based on control by individual states to 200 miles seem unlikely to result in rational fisheries management, it is reasonable to examine next the possibility of developing international management systems similar to the existing IATTC system in the eastern Pacific. Under such a system total fishing effort expended is controlled by setting an upper limit on the total catch of the entire international fleet (i.e., an overall quota). This approach to bringing world tuna resources under international management can be evaluated by first reviewing the history of IATTC management and then considering possible future modifications of this system.

The regulatory program for yellowfin within the CYRA was initiated in 1966, with all participants in the fishery competing on equal footing for shares of the overall quota. When the catch to date plus expected catches of vessels completing unregulated trips plus expected incidental catches equals the quota, unrestricted fishing ceases, and vessels of all flags stop fishing for yellowfin.

From the beginning there were difficulties and differences of opinion among the participating nations, revolving mainly around the issue of catch allocation. Latin American RANs argued that because of their adjacency to the resource, they were entitled to larger shares of the catch than they were taking. In order to secure these larger shares, they requested that national quotas be established and insisted that unless their needs were met, they could no longer participate in the current management program. In opposing this position, the non-RANs, especially the United States, refused to recognize any claims to special allocations based on resource adjacency. They argued that no international precedent existed for granting special rights to migratory pelagic species on the basis of coastal adjacency and that to grant such rights would be contrary to existing international law. In addition, such action would jeopardize their own industries and could

53

unduly influence any decisions concerning jurisdiction over pelagic species that might be reached at a third International Conference on the Law of the Sea.

An obvious threat resulting from this standoff was extension of fisheries jurisdiction to 200 miles with exclusion of foreign fleets. Therefore, as a temporary solution, special allocations were reserved from the quota, and these could be taken in the CYRA after closure to unrestricted fishing. These allocations were based on economic hardship rather than resource adjacency, and they applied to qualifying vessels of any nation. The numbers and types of vessels qualifying for such allocations and the quantities allocated have been renegotiated each year. Under the 1977 regulations there are allocations for small vessels under four hundred tons in carrying capacity and for a certain category of newly constructed vessels suffering unusual hardships. The proportion of the yellowfin quota taken under such special allocations has been steadily increasing, going from nothing in 1968 to about 25–30 percent in 1974.

In addition to establishment of hardship allocations, another major change was taking place in the eastern Pacific tuna fishery. In the early development of the fishery, vessels were free to fish without restraint for tuna wherever they occurred. Under these circumstances, the fishery could properly be called an open-access fishery. After World War II, however, in the late 1940s and early 1950s, the situation began to change. Certain Latin American RANs extended their fisheries jurisdiction to 200 miles during this period, either by extension of their claims to territorial seas or by creation of exclusive fisheries zones. Two countries, Ecuador and Peru, established licensing systems within the 200-mile zones that they claimed, but their enforcement capabilities were limited, and many vessels were able to fish without buying licenses. Because license fees were high, access to tuna resources was somewhat restricted, and the erosion of the open-access concept had definitely begun. The open-access principle was further eroded when Costa Rica and Mexico announced licensing systems in their 200-mile zones.

For years the United States refused to recognize these claims of extended fisheries jurisdiction. Because of this and because of improved Ecuadorian and Peruvian enforcement capability, United States domestic legislation was passed in 1969 creating an insurance fund to compensate owners of seized vessels for the fines that they were forced to pay which, in the case of Ecuador, included the cost of a license. Vessel owners were also partially compensated for lost fishing time resulting from seizure.

To put the situation in quantitative terms, if just Ecuador, Peru, Costa Rica, and Mexico require high license fees to fish within their 200-mile zones, then fleets of other nations face either much higher costs or complete exclusion from fishing grounds that have produced 53 percent of the yellowfin and 74 percent of the skipjack catches taken in the CYRA during the 1970–74 period. Clearly the eastern Pacific fishery can no longer be considered an open-access fishery.

With the steady curtailment of open-access fishing and with increasing shares of the allowable catch set aside for hardship allocations, negotiations each year have tended to become more lengthy and agreement on allocations and the size of the quota more difficult to reach. Under these circumstances, is it reasonable to expect that this same approach or some modification of it can be continued in the future? Three basic courses seem available: Special allocations could be reduced sharply or even eliminated altogether; they could be maintained at approximately their present levels; or they could be increased. Whichever course might be taken, if negotiations broke down and agreement could not be reached, the fishery would revert to some form of control by RANs within their 200-mile coastal zones, or else a new kind of international agreement would have to be found. Thus the question is: How likely is it that overall quota management agreements of the IATTC type, with or without special allocations, can be successfully negotiated in the future?

Suppose first that special allocations are to be sharply reduced and eventually eliminated. Suppose also that any RAN policies that limit access to their 200-mile zones are phased out. Everyone would return to an open-access system similar to the original management system implemented in 1966. An overall catch quota would be established to be taken on a first-come, first-served basis. The season would open on a specified date, vessels of all nations would compete with each other, each taking as large a share of the catch as possible, and when the catch approximated the quota all fishing would cease. In view of RAN attitudes favoring extended jurisdiction and establishment of resource related allocations, it is virtually inconceivable that the RANs would agree to such an arrangement, but if somehow they did, it is probable that the international fleet would grow rapidly. Each nation would want a larger fleet in order to harvest a larger share of the catch as rapidly as possible during the open season. Likewise, within a nation, producers and processors competing for shares of the catch would want more vessels. With fleet growth, pressures to fish all year would become so great that it would be exceedingly difficult to maintain an international conservation program. Even in the unlikely case that the conservation program could be maintained, fleet growth beyond current levels could only result in economic chaos.

Next, suppose that an effort is made to maintain the present system with all special allocations at their current levels. The trend towards establishment of 200-mile fisheries jurisdictions in conjunction with high license fees would have a major impact on the ultimate amount and distribution of fishing effort (see discussion of licensing in Chapter 6). There would still be fierce competition for available resources, and negotiations would intensify. Some nations would want to increase their own allocations, while other nations would strive to reduce allocations generally or at least keep them constant. Pressure to allow fishing on a year-around basis would increase. Under such a pattern of increasing competition and intense negotiation, it is very probable that participating governments would at some point fail to reach agreement on a cooperative conservation program.

Finally, suppose that participating nations want to sustain the present conservation program, but with increased shares of the catch set aside for special allocations. Assuming that this would be possible, it is difficult to imagine significant new allocations being granted for economic hardship reasons. Latin America RANs are increasingly insistent that catch allocation must not be established on the basis of economic hardship, but rather that the issue of their rights to preferential treatment must be directly confronted by basing allocations on the principle of resource adjacency. Adjacency-based yellowfin allocations could result in a substantial portion of the available resource being reserved for RANs. For example, if allocations were set equal to average coastal zone catches by the entire international fleet during the 1970–74 period, 57 percent of the total yellowfin catch would be involved (see Table 2). Presently there are no hardship allocations for skipjack because there is no overall skipjack quota, but if adjacency-based skipjack allocations were established, they could be even larger proportionally than yellowfin allocations because the bulk of the skipjack catch is taken in coastal waters.

It seems inconceivable that the United States and other non-RANs would agree to large adjacency allocations while at the same time being limited in their access to important fishing grounds falling within national 200-mile zones. Instead, they would probably choose to fish as intensively as possible outside 200 miles with consequences which were discussed in the preceding chapter. Some RANs might also be unhappy with adjacency-related allocations. For example, resource-poor nations could receive adjacency allocations smaller than their present economic hardship allocations. Also, assuming that allocations are nontransferable as at present, a nation requires a fleet in order to derive any benefit from an allocation, and some nations might prefer a participant fee system that would allow them to derive benefits regardless of whether or not they have a fleet. Such a system is discussed in the following chapter.

In conclusion, it does not appear that the existing overall quota management system in the eastern Pacific or any modification of it can adequately resolve the catch distribution problem. The RANs, generally speaking, want adjacency-based allocations, and at the same time some are restricting access to the resource by imposing high license fees. The non-RANs would like to do away with allocations altogether and operate the fishery strictly on an open-access, overall quota basis. Both sides are in a strong bargaining position—the RANs because they control access to important parts of the resource, and the non-RANs because their fleets dominate the harvesting sector and they control the markets. If the catch distribution problem cannot be resolved, it hardly seems likely that the related problems of excessive fleet growth and enforcement can be effectively dealt with. Under these circumstances, it appears that the continued success of the present IATTC management program is increasingly jeopardized.

8   Open Access with Participant Fees and Resource Adjacency Allocations (PAQ Management)

Several alternatives for future management of tuna resources have been considered under the assumption that coastal states will extend their fisheries jurisdiction to 200 miles on a nearly universal basis by 1980. Systems involving control by individual coastal states within their own 200-mile juridical zones were considered first and found to be wanting (Chapter 6). Specific shortcomings depended to some extent on the management approach under consideration, but all approaches were plagued by a failure to adequately resolve the catch allocation problem. RANs wishing to harvest tuna, say in quantities equivalent to the average amounts taken by the international fleet from their coastal waters, cannot possibly do so year after year if they are restricted to fishing only in their own juridical zones because of the highly variable migratory behavior of tuna. Non-RANs, if excluded from coastal zones by high license fees or other RAN policies, would concentrate their enormous fishing power beyond 200 miles, possibly overexploiting the stock there. Under such circumstances, it would be virtually impossible to achieve agreement on a conservation program covering the entire resource. Obviously some type of international management system is required, so extensions and modifications of the eastern Pacific IATTC system in effect in 1977 were considered next (Chapter 7). It was concluded that this approach could not provide a long-term solution to the catch allocation problem either. RANs would insist on catch allocations in recognition of their adjacency to the resources, but non-RANs would refuse to grant such a concession as long as open access to important fishing areas was limited by various RAN licensing schemes. Under these pressures it is unlikely that the IATTC conservation program can survive for long. And if conservation and catch distribution problems cannot be resolved, neither will it be possible to deal with fleet growth and enforcement problems.

From the analysis thus far, five essential features can be deduced which should be incorporated into an international tuna management system if it is to succeed:

1. The fishery should operate under a conservation and management program that applies to the exploited resource over its entire range, and this program should include species taken in association with tuna such as porpoise.
2. Some form of recognition should be given to RAN claims that they are entitled to special allocations or other kinds of special treatment because of their adjacency to the resource.
3. All participants in the fishery, RANs and non-RANs alike, should be guaranteed open access to all important fishing grounds, including those within RAN coastal zones.
4. Effective means should be developed for limiting fleet expansion.
5. A capability should exist for providing uniform enforcement of conservation regulations.

An approach to management incorporating these five features will now be described in detail (see Chapter 12 for discussion of porpoise conservation). Management would be based on open access to all resources and would be administered by an international agency that sets overall catch quotas, issues international licenses, partially allocates the catch to coastal nations, collects participant fees based on catches, redistributes resulting proceeds among nations, and provides for enforcement and control of fleet size. The distinguishing feature of such a management system is that the overall quota would in some way be partially allocated among RANs in recognition of their adjacency to the resources. Hence the term "partially allocated quota" management, or simply PAQ management, will be used in referring to this type of management.

It is believed that some version of PAQ management could be successfully implemented in the eastern Pacific. No nation or segment of industry would be entirely satisfied with the system ultimately evolved through negotiation because no participant would get all that it would ideally like to have. Nevertheless the most essential needs of all participants could be recognized and provided for to some extent, and it should be possible to maintain productivity at near maximum levels. At best, any PAQ system that is adopted will represent a compromise among conflicting philosophies, but as has been stressed, cooperation and compromise are essential if the critical problems presently facing the tuna industry are to be resolved, either in the eastern Pacific or on a worldwide basis.

Partial Catch Allocation and Resource Adjacency

In theory, anywhere from 0 to 100 percent of the total catch could be allocated among nations, and many possible bases exist for determining the size of national allocations, some of which are discussed in Chapter 10. However, as has been noted earlier, many RANs strongly believe that allocations should be based on resource adjacency. Suppose, therefore, that there is agreement in principle that establishment of a viable long-term management program for tuna will necessitate institution of special allocations for RANs that are related in some way to their adjacency to tuna resources. The question arises immediately as to just how such allocations should be determined. In a well-developed fishery, such as that for yellowfin in the eastern Pacific, the historic geographic distribution of catches reflects the distribution of harvestable resources in RAN coastal waters, and it seems reasonable that yellowfin allocations should be related in some way to this distribution rather than to some other measure of adjacency such as density or abundance. This seems appropriate because in exploiting a fishery resource the simplest and most concrete measure of success is the magnitude of the catch. Also catches can be directly and accurately monitored, whereas the density or abundance of stocks can only be inferred by indirect methods (such as monitoring of catch per unit of effort), which are subject to considerable statistical error and fraught with sources of possibly serious bias. Allocating resources on the basis of catches in adjacent waters also seems consistent with the philosophy that nations should have the right to exercise control over and participate in the harvesting of migratory species in their coastal waters, but should not claim ownership of the fish themselves, for tuna can move freely through the juridical zones of many nations and to the high seas beyond these zones. By contrast, allocations based on indexes or estimates of density or abundance would seem to imply ownership or property rights, an undesirable connotation. A final reason for basing adjacency allocations on actual catches is that they are unambiguous, easy to calculate, and readily understood.

In determining RAN allocations based on catches in adjacent waters, two conflicting philosophies must be considered. On the one hand, Latin American RANs with significant resources in their 200-mile zones would naturally want the magnitude of these resources to weigh heavily in making allocations. On the other hand, the United States, which developed the eastern Pacific fishery and has by far the largest fleet, would naturally want more weight assigned to historic RAN catches and present RAN harvesting capability. Both of these viewpoints will be considered.

The view that allocations should be based on resource adjacency would receive maximum recognition if each coastal state were allocated a share of the yellowfin catch equivalent to the average catch taken in its coastal waters by the entire international fleet in recent years. One could not be more generous, for there does not appear to be any basis whatsoever for a RAN to claim special rights to a larger amount. The remainder of the overall yellowfin quota, an amount equal to the average catch made on the high seas beyond 200 miles by the entire international fleet, would be unallocated.

By contrast, maximum emphasis would be placed on historical RAN catches and present RAN harvesting capability if each coastal state were allocated a share of the yellowfin catch equivalent to the catch taken in its own coastal zone by its own fleet in recent years. It is difficult to imagine any rationale for reducing allocations below this level unless one argues that there should be no adjacency-related allocations at all. This approach would leave a much larger share of the overall yellowfin quota unallocated.

In the case of skipjack, there is no present basis for imposing an overall quota on the eastern Pacific catch. In fact, present policy is just the opposite: Harvesters are encouraged to maximize their catches whenever and wherever skipjack become available. Because fishing for skipjack is unrestricted and because their availability varies greatly from year to year, catches also vary greatly. Under these conditions it does not presently seem reasonable to create allocations for skipjack. Instead, another means must be found if RAN claims to preferential treatment based on their resource adjacency are to be recognized. A participant fee system that would provide such a means will be discussed later in this chapter. If restriction of skipjack catches eventually becomes necessary, then RAN allocations might be established, possibly on a basis other than that used in the case of yellowfin. Skipjack management is discussed in greater detail in Appendix II.

Like skipjack, other species taken in the eastern Pacific fishery such as bigeye tuna, bluefin tuna, and bonito would not be allocated initially, but they could be subject to participant fees. In the case of billfish taken by longliners, there are special problems that are discussed in Chapter 12.

In allocating catch shares, the question of whether or not they should be transferable is an important one, especially if they are substantially larger than present RAN catches. If a nation is free to trade, sell, or otherwise dispose of all or part of its allocation, then what has been created is a property right in terms of fish. On the other hand, if allocations are nontransferable, then what has been created is the privilege of guaranteed access to a portion of the resource. If allocations are treated as transferable property rights, then the benefit that a RAN might derive from any unutilized portion of its allocation would depend strictly on what kind of deal it could negotiate for itself. For at least a few years such transactions would generally involve transfers from RANs to non-RANs, thus somewhat mitigating the impact of allocation establishment on non-RAN operations. Similarly, if allocations are nontransferable guarantees of access, then unutilized portions would be added to the unallocated share of the overall quota and harvested primarily by non-RAN fleets. In this latter case, some provision should be made to compensate RANs that forego their privilege to fully harvest their allocations. Once again, a system of participant fees would be a means for providing such compensation. Although either approach would be potentially workable, it will be assumed throughout the remainder of this discussion of PAQ management that allocations are nontransferable guarantees of access rather than transferable property rights, unless otherwise stated.

In creating allocations, no matter what their magnitude, there would be no intent to limit the RAN catches to any specific level. An individual RAN could be a nonparticipant in the fishery, could harvest less than its allocation, or could exceed its allocation. If a RAN wanted to harvest at levels exceeding its allocation, then it would have to do so by competing with the international fleet for the unallocated share. The terms under which this competition would take place are very important to RANs and non-RANs alike and are discussed in Appendix I, which deals with procedures for closure of the fishery.

To illustrate how PAQ management might operate, detailed, step-by-step examples will be developed. It is stressed that these are purely hypothetical examples, not proposals or recommendations. They are intended to define a range of possibilities and to bring out certain problems that must be considered in establishing PAQ management. Two types of allocations to RANs corresponding to the contrasting Latin American and non-RAN viewpoints mentioned earlier will be considered: (1) allocations equivalent to recent average catches by the entire international fleet in national 200-mile zones; and (2) allocations equivalent to recent catches by coastal state fleets within their own 200-mile zones. In the former case maximum emphasis is placed on resource adjacency resulting in high allocations. In the latter case historical RAN catches and harvesting capability receive maximum emphasis resulting in low allocations. For the sake of clarity, the examples treat only yellowfin and skipjack, but the concepts illustrated could apply equally to other species.

The average annual yellowfin catch from the entire eastern Pacific during the 1970–74 period amounts to about 192,000 tons. In the high allocation example, adjacency-based yellowfin allocations are determined by the average catches of the entire international fleet within 200-mile coastal zones during this 1970–74 period. These figures are contained in Table 2, but for convenience they are extracted from that table and presented again in the first part of Table 3. Three nations have by far the largest yellowfin allocations under this scheme: Mexico, Costa Rica, and Ecuador. Then come three nations with moderate allocations: Panama, Peru, and Colombia. The remaining RANs trail with small allocations. Although the United States is not considered a RAN, it does receive a tiny allocation under this scheme. Table 3 also shows average skipjack catches. Although no skipjack allocations would be created initially, this information is included to facilitate later illustration of a participant fee system and other aspects of PAQ management. The largest skipjack catches were made off Ecuador, Mexico, and Costa Rica, the same three nations that lead in the case of yellowfin. It is likely that the total skipjack catch can be increased beyond the 70,000 tons taken in the 1970–74 period. For example, during the 1975–77 period skipjack catches averaged 124,000 tons per year.

The most important single fact concerning this high allocation scheme is that 57 percent of the yellowfin catch would be allocated. This figure can be compared to 1970–74 average non-RAN catches which amounted to about 82 percent of the yellowfin harvest, the United States share being about 74 percent (70 percent for the 1975–77 period). Thus, if this scheme were adopted and RANs took their entire allocations, non-RAN yellowfin catches would decline by

TABLE 3
MAXIMAL AND MINIMAL CRITERIA FOR PARTIALLY ALLOCATING AN OVERALL YELLOWFIN QUOTA IN THE EASTERN PACIFIC

PART 1: Average annual catches (in short tons) within national 200-mile zones by the entire international fleet, 1970–74

Nation	Yellowfin (high allocations) Proportion	Amount	Skipjack (not initially allocated) Proportion	Amount	Both species combined Proportion	Amount
Colombia	.0317	6,088	.0695	4,848	.0417	10,936
Costa Rica	.1181	22,710	.1494	10,420	.1264	33,130
Ecuador	.0847	16,293	.2571	17,934	.1306	34,227
El Salvador	.0187	3,593	.0166	1,158	.0181	4,751
France (Clipperton)	.0197	3,784	.0037	256	.0154	4,040
Guatemala	.0213	4,098	.0147	1,022	.0195	5,120
Mexico	.1872	36,011	.2037	14,209	.1916	50,220
Nicaragua	.0087	1,671	.0124	867	.0097	2,538
Panama	.0417	8,023	.0300	2,092	.0386	10,115
Peru	.0341	6,553	.1002	6,991	.0517	13,544
U.S.A.	.00004	9	.0001	7	.0001	16
Within 200 miles	.5658	108,832	.8575	59,806	.6434	168,637
Beyond 200 miles	.4342	83,522	.1425	9,941	.3566	93,461
Total average catch	1.0000	192,353	1.0000	69,746	1.0000	262,099

PART 2: Maximum annual catches (in short tons) by a nation's own fleet from within its own 200-mile zone, 1970–74

Nation	Yellowfin (low allocations)		Skipjack (not initially allocated)		Both species combined	
	Proportion	Amount	Proportion	Amount	Proportion	Amount
Colombia	–	0	–	0	–	0
Costa Rica	.0043	818	.0020	141	.0037	959
Ecuador	.0469	9,020	.1910	13,324	.0853	22,344
El Salvador	–	0	–	0	–	0
France (Clipperton)	.0002	41	–	0	.0002	41
Guatemala	–	0	–	0	–	0
Mexico	.0417	8,012	.0457	3,189	.0427	11,201
Nicaragua	–	0	–	0	–	0
Panama	.0069	1,325	.0100	694	.0077	2,019
Peru	.0081	1,554	.0279	1,948	.0134	3,502
U.S.A.	.0001	25	.0005	37	.0002	62
Subtotal	.1081	20,795	.2772	19,333	.1531	40,128
Remainder of average catch	.8919	171,558	.7228	50,413	.8469	221,971
Total average catch	1.0000	192,353	1.0000	69,746	1.0000	262,099

NOTE: In part 1, high allocations are based on average annual catches taken within 200 miles of coastal nations by the entire international fleet. In part 2, low allocations are based on maximum annual catches by each nation's own fleet from within its own 200-mile zone. Sums of individual average catches and proportions may not agree exactly with totals shown because of rounding.

47 percent and would represent only 43 percent of the total yellowfin catch. Such a reduction would have drastic consequences for non-RAN fleets. At least initially, however, substantial portions of the RAN quotas would be unutilized and therefore available to non-RAN fleets. Also, non-RAN fleets could probably increase their skipjack catches in the eastern Pacific. The manner of RAN fleet development (i.e., either by flag changes or by new construction) would also have a bearing on the success of non-RAN fleets remaining in the eastern Pacific, as will be illustrated in Appendix III.

In the case of low allocations it would not be reasonable to use average catches by coastal states in their own coastal zones during the 1970–74 period because some RAN fleets were developing during these years. Nor would it be reasonable to base allocations on catches in any single recent year because of the great annual variations in tuna distribution and abundance. To avoid these problems and to treat RANs as favorably as possible under the low allocation approach, each nation's allocation could be set equal to the maximum annual catch taken by its own fleet within its own 200-mile zone during the 1970–74 period. Adopting this approach would result in the yellowfin allocations shown in the second part of Table 3. Once again, comparable figures are also given for skipjack.

The most striking aspect of the low allocations is that in total they comprise only 11 percent of the yellowfin harvest. This is a very small amount, especially when compared to the high allocations, which represent 57 percent of the total yellowfin catch. Ecuador is the only country whose low yellowfin allocation would be anywhere close to its high allocation, and the same is true of skipjack although this species would not be allocated initially. Allocations for all other nations either border on insignificance or are nonexistent. There are several reasons for these low figures. The coastal resources of some nations are limited so that even a nation like Panama, which had a reasonably large fleet during the 1970–74 period, has a small allocation. Other nations had either small fleets or no fleets at all. Some of these nations such as Nicaragua and Guatemala are also resource limited, but Costa Rica receives a very low allocation even with its large resources. Finally, a nation like Mexico can have both substantial resources and a fairly good-sized fleet, but if that fleet takes significant catches outside its own 200-mile zone, it will receive correspondingly smaller allocations. Only Ecuador had a fairly well-developed fleet fishing on substantial resources falling primarily within its own coastal zone.

As already noted, a participant fee system such as that to be described later in this chapter could enable RANs to derive benefits from coastal zone catches of skipjack, an initially unallocated species, and also from unutilized portions of yellowfin allocations. A participant fee system could also provide benefits to RANs based on the difference between their allocations and catches made by the entire international fleet in their coastal zones. This difference would be zero in the case of high allocations and would be maximized in the case of low allocations.

Whatever basis is ultimately chosen for allocations, it is important to bear in mind that the distribution and local abundance of tuna vary from year to year. Changes in skipjack catches off the coasts of Peru and Ecuador provide a good example. In the 1960s these catches were very substantial (see Table 2), but they have fallen off sharply in the 1970s as the area of high skipjack catches has shifted north to off Central America, apparently owing to changing environmental conditions. Other changes seem likely in the future. Hence, it might be desirable to update national allocations from time to time so that they more nearly reflect recent conditions. This could be done at specified intervals, basing new allocations on data from a specified number of years; or it might be done annually, using a moving average; or it could be accomplished by some other method. It would be up to the nations themselves to decide how to deal with this matter. It is of interest, however, to note that if allocations based on each nation's catch in its own coastal zone are to be updated, then nations adjacent to large resources such as Mexico, Costa Rica, and Ecuador would have a considerable incentive to develop their fleets. A larger fleet could take a larger catch in its own coastal zone, which would lead to increased allocations, and differences between high and low allocations could be reduced.

Open-Access Fishing and International Licensing

It is unlikely that non-RANs would agree to cooperate in any management program involving reduction of their catches through creation of yellowfin allocations unless they receive the benefit of some major RAN concession in return. Because it is important to fleets of all nations to fish for tuna wherever they occur and regardless of whose juridical zone they may be migrating through, a logical RAN concession would be to allow open access to fishing grounds within their juridical zones for vessels of all nations cooperating in the management program. This would enable fleets to fish wherever fish were concentrated without buying a series of expensive fishing licenses. In some years this might be inshore off Central America, in other years inshore off Peru and Ecuador, in still other years offshore from Mexico, and so on. With open access, a major problem plaguing the present eastern Pacific IATTC management program would be overcome, and there would be a reasonable hope for lasting management arrangements.

It should be stressed that non-RANs are not the only nations requiring open access to all major fishing grounds. Catches in the coastal waters of every RAN show a high degree of variability around the average catch (Table 2). If negotiations result in relatively high RAN yellowfin allocations (perhaps approaching those shown in the first part of Table 3), it would be unusual for a RAN to be able to harvest its full allocation from its own coastal waters in any given year. In some years, the availability of yellowfin in a nation's coastal zone would be so insufficient that it could not fill its allocation there no matter what the size of its fleet. In other cases, a nation's fleet would be too small to harvest its full allocation during a limited period of availability, even if large resources became available. Hence open access to all fishing grounds would be essential for RANs if relatively high yellowfin allocations are adopted.

The situation is similar both in the case of low yellowfin allocations (second part of Table 3) and in the case of skipjack, for which there would initially be no allocations. If a RAN wished to harvest more than its small yellowfin allocation and also to have adequate access to skipjack resources, its vessels would need to fish on major concentrations wherever they occured. If they were not able to do so, their catch expectations could not be realized.

If RAN and non-RAN vessels are to fish on an open-access basis, foreign flag vessels must be given permission to enter the coastal or juridical zones of other states. In the past the customary way of allowing access has been for individual states to require foreign flag vessels to purchase fishing licenses or permits. In most cases such licenses have been extremely expensive, and the many drawbacks of national licensing have already been discussed in Chapter 6. A better approach would be to provide international licenses for flag vessels of all nations willing to cooperate in an international resource management and conservation program. A single international license would entitle its holder to fish anywhere in the eastern Pacific to within, say, 12 miles of the coastline of any participating nation. Thus, both RAN and non-RAN fleets would have access to major concentrations of fish, whether they occur inside juridical zones or on the high seas beyond such zones.

This international licensing scheme would operate in conjunction with an international system of participant fees that will be described in the next section. Very briefly, however, vessels of nations cooperating in the management program and paying participant fees would be provided with an international fishing license, and a portion of the participant fees would be distributed to RANs. In a sense, these payments would compensate them for such licensing. A significant advantage of international licensing is that vessels of noncooperating nations (both RANs and non-RANs) would be entirely excluded from fishing in the 200-mile juridical zones of participating nations. This would sharply restrict their access to the tuna resources and should act as a strong deterrent to noncooperation. Participants could also cooperate in imposing other sanctions against nonparticipant nations. For example, landing or transshipment of fish could be disallowed, use of port facilities banned, or import embargoes enforced against both unprocessed and processed products. It would be desirable, however, to avoid such sanctions if at all possible.

International licensing of foreign flag vessels to fish for tuna inside a nation's juridical zone to within 12 miles of shore raises the question of whether or not the sovereign rights of coastal nations would be infringed upon. Essentially the same sovereignty question would also arise in connection with international monitoring of fishing activities and international enforcement of conservation regulations. There are three basic reasons why this sovereignty question should not be unduly difficult to resolve.

First, it can be persuasively argued that in agreeing to a cooperative international management program operating within national juridical zones as well as on the high seas, all involved nations would, in effect, be recognizing one another's sovereign rights. After all, a nation must have the authority to control fishing within its juridical zone before it can agree to delegate that authority to an international agency for the sake of rational resource management which provides

TABLE 4. ANNUAL YELLOWFIN AND SKIPJACK CATCHES TAKEN WITHIN INTERNATIONAL

Yellowfin catch within 12 miles of --	1974	1973	1972	1971	1970	1969	1968	1967
Chile	0	0	0	0	0	0	0	0
Colombia	249	1,626	263	10	68	466	54	188
Costa Rica	472	2,750	254	1,345	535	49	140	8
Ecuador	1,116	2,058	1,073	2,022	1,754	859	1,949	1,761
El Salvador	9	0	9	0	0	2	4	0
France (Clipperton)	274	255	31	33	91	54	0	0
Guatemala	50	14	4	9	2	19	28	7
Mexico	4,583	4,193	2,303	3,497	7,902	4,507	4,155	6,425
Nicaragua	54	23	7	85	9	6	118	3
Panama	49	2,650	589	186	65	12	11	25
Peru	975	2,400	798	1,352	1,341	502	612	1,599
U.S.A.	2	0	0	5	4	0	2	0
Total within 12 miles	7,833	15,969	5,331	8,544	11,771	6,476	7,073	10,016

Skipjack catch within 12 miles of --	1974	1973	1972	1971	1970	1969	1968	1967
Chile	0	0	0	0	0	0	0	0
Colombia	618	1,902	254	206	11	614	330	266
Costa Rica	425	272	13	1,295	56	0	32	2
Ecuador	616	674	1,190	4,328	2,200	5,699	5,079	12,054
El Salvador	4	0	0	0	0	0	0	0
France (Clipperton)	0	0	0	0	34	9	0	0
Guatemala	32	0	1	4	0	0	0	0
Mexico	697	546	1,435	1,560	4,445	1,638	1,372	2,120
Nicaragua	157	0	0	177	22	0	9	1
Panama	6	130	48	500	0	0	1	14
Peru	399	1,476	452	1,614	1,119	2,312	2,094	3,246
U.S.A.	0	0	0	0	22	1	2	4
Total within 12 miles	2,954	5,000	3,393	9,684	7,909	10,273	8,919	17,707

Combined yellowfin & skipjack catch within 12 miles of --	1974	1973	1972	1971	1970	1969	1968	1967
Chile	0	0	0	0	0	0	0	0
Colombia	867	3,528	517	216	79	1,080	384	454
Costa Rica	897	3,022	267	2,640	591	49	172	10
Ecuador	1,732	2,732	2,263	6,350	3,954	6,558	7,028	13,815
El Salvador	13	0	9	0	0	2	4	0
France (Clipperton)	274	255	31	33	125	63	0	0
Guatemala	82	14	5	13	2	19	28	7
Mexico	5,280	4,739	3,738	5,057	12,347	6,145	5,527	8,545
Nicaragua	211	23	7	262	31	6	127	4
Panama	55	2,780	637	686	65	12	12	39
Peru	1,374	3,876	1,250	2,966	2,460	2,814	2,706	4,845
U.S.A.	2	0	0	5	26	1	4	4
Total within 12 miles	10,787	20,969	8,724	18,228	19,680	16,749	15,992	27,723

NOTE: Sums of individual average catches may not agree exactly with totals shown because

12 MILES OF COASTAL NATIONS IN THE EASTERN PACIFIC BY THE ENTIRE FLEET (Short tons)

1966	1965	1964	1963	1962	1961	Average catches 1970-1974	1966-1974	1961-1974
41	0	108	61	419	15	0	5	46
434	11	111	661	3,339	180	443	373	547
187	202	295	419	1,019	837	1,071	638	608
3,496	1,381	2,182	1,817	2,406	315	1,605	1,788	1,728
2	5	2	8	6	81	4	3	9
16	4	76	11	251	282	137	84	98
12	27	3	12	15	394	16	16	43
3,632	5,325	4,414	6,388	5,654	3,443	4,496	4,577	4,744
12	58	72	162	152	611	36	35	98
158	33	9	175	243	343	708	416	325
2,478	1,145	2,186	1,581	2,736	381	1,373	1,340	1,435
0	0	0	7	0	1	2	1	2
10,468	8,191	9,458	11,302	16,240	6,883	9,890	9,276	9,683

1966	1965	1964	1963	1962	1961	Average catches 1970-1974	1966-1974	1961-1974
6	21	41	45	20	45	0	1	13
490	147	435	2,991	3,165	60	598	521	821
54	295	503	648	876	759	412	239	374
7,551	13,503	4,389	14,273	10,546	8,636	1,802	4,377	6,481
0	0	3	29	19	92	1	0	11
0	0	0	0	0	0	7	5	3
4	1	0	0	1	28	7	5	5
787	2,754	2,069	1,929	306	820	1,737	1,622	1,606
3	103	265	705	793	1,255	71	41	249
298	15	7	66	75	244	137	111	100
3,979	2,937	1,884	3,095	2,237	2,581	1,012	1,855	2,102
79	5	0	116	0	1	4	12	16
13,251	19,781	9,596	23,897	18,038	14,521	5,788	8,788	11,780

1966	1965	1964	1963	1962	1961	Average catches 1970-1974	1966-1974	1961-1974
47	21	149	106	439	60	0	5	59
924	158	546	3,652	6,504	240	1,041	894	1,368
241	497	798	1,067	1,895	1,596	1,483	877	982
11,047	14,884	6,571	16,090	12,952	8,951	3,406	6,164	8,209
2	5	5	37	25	173	4	3	20
16	4	76	11	251	282	144	89	102
16	28	3	12	16	422	23	21	48
4,419	8,079	6,483	8,317	5,960	4,263	6,232	6,200	6,350
15	161	337	867	945	1,866	107	76	347
456	48	16	241	318	587	845	527	425
6,457	4,082	4,070	4,676	4,973	2,962	2,385	3,194	3,537
79	5	0	123	0	2	7	13	18
23,719	27,972	19,054	35,199	34,278	21,404	15,678	18,063	21,463

of rounding.

long-term benefits for all who utilize it. From this point of view, cooperation and mutual understanding of commonly shared problems could prove to be a more effective means for securing international recognition of sovereign rights than confrontation, vessel seizures, economic retaliation, defiance of conservation regulations, and the like.

Second, many nations are claiming the right to control all fishing activities within a zone of economic influence extending out to 200 miles from a much narrower territorial sea, frequently 12 miles in width. Agreeing to international control of tuna fishing within an economic influence zone is not quite the same as doing so within a much narrower territorial sea immediately adjacent to a nation's shore. The concept of an economic influence zone falls between the concept of a territorial sea where traditionally sovereignty has been absolute and the concept of the high seas where traditionally no sovereign rights have been recognized. Agreement to international management of a highly migratory resource exploited by a mobile international fleet would seem to be in keeping with the economic influence zone concept.

Finally, suppose that nations, agree to accept international management within economic influence zones, but continue to exercise absolute and undiluted sovereignty over a 12-mile territorial sea. Would exclusion of fisheries within the 12-mile zone unduly compromise an international management program? To answer this question, at least for the eastern Pacific, Table 4 was prepared, showing catches of yellowfin and skipjack tuna within 12 miles of various nations during the 1961–74 period. Generally speaking, catches within 12 miles are relatively small, the principal exception being skipjack catches off Ecuador, where in certain years rather substantial quantities are taken. Considering only the 1970–74 period, the greatest catches by a substantial margin were made within 12 miles of Mexico, with Ecuador and Peru following. The position of Mexico and Ecuador is reversed, however, if the entire 1961–74 period is considered. In any event, catches are small compared to total catches, in recent years representing only 5 percent and 8 percent respectively of eastern Pacific yellowfin and skipjack production. On the basis of these figures, there is no particular reason or need for open-access fishing in the 12-mile zone. Thus, 12-mile territorial zones could be closed to foreign flag fishing without endangering achievement of an international management program. However, yellowfin catches made within 12 miles by a flag nation's own vessels (or any other vessels that it may allow to fish there) should be counted both in determining catches under national allocations and as part of the overall quota, because these fish are part of the migratory intermingling eastern Pacific stocks under management. Licensing of baitboats to capture bait fish within 12-mile territorial sea zones would, of course, be at the discretion of coastal states.

The possibility of periodically revising RAN allocations was mentioned earlier. With open access to all important fishing grounds through international licensing, catches should reasonably reflect the distribution and availability of yellowfin in the eastern Pacific. Hence, catches by either the entire international fleet or by the fleets of individual coastal states could be used as a basis for establishment of new allocations just as in the examples of Table 3. There are possible problems, however. Nations could conceivably concentrate their fishing in certain areas and avoid other areas in order to try to bring about allocations more favorable to themselves. For example, non-RANs might concentrate their fishing beyond 200 miles as much as possible in order to generally reduce RAN allocations. However, RANs especially those with larger fleets, might try to maximize catches in their own waters to increase their own allocations. Because these are offsetting strategies, biased distribution of national fishing effort might not be much of a problem. Nevertheless some special consideration might be necessary for RANs that choose to derive benefits from participant fee disbursements rather than by developing fleets of their own. Both of the fishing strategies just mentioned would tend to inequitably reduce the allocations of such RANs. A related problem would be actual falsification of data concerning location of catches in order to gain an advantage if and when allocations are revised. This could be dealt with effectively by a sophisticated system for monitoring catches and landings.

Collection and Redistribution of Participant Fees

In discussing PAQ management, recognition of RAN claims to some form of special treatment based on their resource adjacency has been considered. It was noted earlier in this chapter that such recognition should probably be in some way related to the distribution of catches within their 200-mile zones. At one logical extreme, high allocations could be based on coastal zone catches by the entire international fleet. At the other logical extreme, low allocations could be based on catches by the individual coastal states within their own zones. Because any agreement will have to come out of give-and-take negotiations, allocations at some intermediate level are perhaps most likely.

No matter what the final basis for allocations, they cannot by themselves fully resolve the problem of providing preferential treatment for coastal states. Several situations clearly illustrating this fact have already been noted. For example, if high yellowfin allocations were to be adopted, RANs that refrain from fully utilizing their allocations should be compensated in some way. Also, in the case of low or intermediate allocations, a nation would want somehow to derive benefits based on the difference between its total catch and the catch taken by the entire international fleet in its waters. This is especially true in the case of nations with underdeveloped fleets that are adjacent to rich tuna waters (e.g., Costa Rica). RANs would also want to derive benefits based on catches of skipjack, a species which would be unallocated initially. Obviously some system for redistributing benefits derived from the fishery among harvesting nations and RANs is required, and it should be fully compatible with the system for determining allocations. If a satisfactory mechanism cannot be found for redistributing benefits, any management system will be unstable and plagued with unresolvable problems.

Assuming that allocations are nontransferable guarantees of access to shares of the resource, a redistribution system should provide benefits to (or transfer benefits from) each RAN that are directly related to the difference between the catch of the entire international fleet within the nation's 200-mile zone and the total eastern Pacific catch made by the nation's own fleet. If allocations are transferable property rights, then procedures for redistributing benefits must be modified in an appropriate fashion, but the underlying principle of deriving benefits based on the magnitude of coastal zone resources would still hold. For the time being, however, allocations will be assumed to be guarantees of access. Transferable allocations will be briefly considered at the end of this section.

Fortunately, a properly structured system of participant fees could provide an appropriate mechanism for redistribution of benefits. In addition to meeting this fundamental need, participant fees could also provide a source of funds for support of international management. Because a portion of the participant fees collected from harvesters would be set aside for management purposes, those who derive benefit from rational international management of the fishery would be the ones who in turn make management possible, as seems only fitting. This participant fee system will be described using the example of the eastern Pacific yellowfin and skipjack fishery. However, other species of tuna, tuna-like fish, and billfish could also be subject to participant fees, and such a system could operate in any ocean area, or even globally.

Participant fees, which can be thought of as a kind of catch tax, would be imposed on all harvesters, RANs and non-RANs alike. A certain fixed amount would be charged for each ton of tuna captured in the eastern Pacific no matter where caught, with all proceeds going to the international management body for redistribution. Fee levels would have to be mutually agreed upon by the nations involved. On the one hand, they should be set high enough to produce funds sufficient to support management and provide for reasonable disbursements to RANs. On the other hand, the requirement that harvesting activities must be carried out on a profitable basis puts an upper limit on the fee level. Also, fees should be established on a species-by-species basis because of variations in product value.

The international body would establish a system for monitoring all landings to determine the participant fees to be collected. Each harvesting nation would decide for itself how to pay the fees. For example, some nations might require boat owners to pay when they land their catches, while others might subsidize their fishing fleets by paying with funds from the national treasury. It is important to emphasize that because the same fees would apply over the entire eastern Pacific fishery, it would make no difference in collecting the fees whether a boat fished in its own coastal zone, in the coastal zone of another nation, or on the high seas beyond 200 miles.

Redistribution of participant fees would be carried out on an annual basis and can be envisioned as a three-step process. First, a portion would be allocated for operation of the management body. This amount would have to be sufficient to cover all costs of management including scientific studies, monitoring of fishing activities (collection of data on catch, effort, size composition, fleet composition, etc.), enforcement and surveillance, collection and disbursement of participant fees, issuance of international licenses, and general administration. Second, a portion of the remaining funds representing catches made in RAN coastal zones would be divided up among RANs on the basis of resource adjacency. The final portion, representing catches made on the high seas beyond 200 miles, would then be distributed. Alternatives for this final stage will be considered after discussion of adjacency-based disbursements to RANs.

Resource-adjacency payments to RANs would be distributed on the basis of recent average catches made by the entire international fleet within the 200-mile zones of the various RANs. Neither the level of RAN allocations nor the magnitude of RAN catches would have any effect on this stage of the redistribution process. It is important to see clearly how this procedure is consistent with the fundamental principle that benefits should accrue to a RAN in relation to the difference between the catch in its waters by the entire international fleet and its own total eastern Pacific catch. Suppose a RAN harvests and pays fees on a catch equivalent to the recent average international catch in its own coastal zone. It would receive adjacency-based disbursements exactly equal to what it had paid in less its fair share of the total cost for administration and management of the resources. This is as it should be because the difference referred to above would in this case be zero. If the RAN catch was less than the recent average international catch, the difference would be in its favor, and it would receive back more than it had paid in after adjusting for management deductions. If its catch exceeded the recent average international catch, it would receive less than it had paid in. Non-RANs would pay fees, but would not receive adjacency-based disbursements. A RAN that did not participate in the fishery naturally would not pay any fees. It would, however, receive its adjacency-based disbursement from the fees collected. Such a nation would also, in effect, be paying its fair share for maintenance of the resource because funds for management are deducted before redistribution of participant fees. This means that it would receive less than it would have received if distributions had been made prior to deducting for management. Thus, all nations participating in the fishery as well as all nonparticipating RANs would contribute to the proper management of the resources. This seems just because all derive benefits from well-managed resources.

As an alternative to basing disbursements on recent average international catches, it would be possible to disburse payments to RANs on the basis of annual international catches made each year in their coastal zones. The obvious drawback to this approach is the extreme variability from year to year in the coastal zone catches (Table 2). It would be a feast or famine situation, and in the long run total amounts received by individual RANs would not differ significantly from the amounts they would receive on a much less variable basis if payments were based on reasonably long-term average catches. If average catches are used, the desirability of updating the averages from time to time must be considered, just as in the case of allocations. If the averages are to be updated, it is worth noting that any changes in fishing strategy by either RANs or non-RANs intended to increase allocations would also affect updated participant fee disbursements. Total adjacency disbursements and the relative shares of individual RANs could both be affected. However, RAN and non-RAN strategies should tend to be offsetting, and the problem does not seem very serious.

Alternatives exist for disbursing the final portion of the participant fee receipts representing high-seas catches made beyond 200 miles. For example, non-RAN participants could argue that if RANs receive disbursements based on the total catch made in their coastal zones, then they should not participate in disbursements representing high-seas catches (assuming that RAN catches do not exceed catches in their coastal zones). If RANs have received comparatively large allocations, perhaps they might be willing to accept this philosophy. If they did, the portion of the participant fees attributable to the high-seas catch beyond 200 miles could be distributed among non-RANs (and RANs whose catches exceed their allocations) on the basis of their catches (and RAN catches in excess of allocations) in the eastern Pacific, regardless of where they are made.

The RANs might well reject the above position, however, arguing that high-seas catches from an international resource represent a part of the common heritage of mankind (including themselves) to share in the wealth of the sea. Carried to a logical extreme, this position implies that the high-seas portion should be distributed among all nations of the world. But this seems impractical and questionable in terms of equity. For example, suppose that disbursements were to be made on the basis of population. With at most a few million dollars being distributed, the amount that any nation would receive would be trivial to the point of hardly being worth the administrative costs of handling the transactions. Furthermore, it would seem to be inequitable for nations that pay participant fees and support management not to be primary beneficiaries of high-seas disbursements. A more realistic approach might be to distribute the high-seas portion to harvesting nations (both RANs and non-RANs) on the basis of their actual catches no matter where in the eastern Pacific they are made. Another alternative would be to use the money in a way that would benefit all of mankind (e.g., support basic research in areas such as marine ecology or marine pollution). If such a philanthropic approach were adopted, emphasis could be placed on projects whose primary direct value was to nations participating in the fishery (e.g., resolution of the tuna-porpoise problem, gear development, etc.) Still another possibility would be to use the high-seas portion to provide financial incentives for RANs to forego harvesting their catch allocations so that these shares would be available to non-RAN fleets. This concept is explored further in Appendix III.

In disbursing the participant fee fund, the international management body would make payments directly to governments of recipient nations. It would be up to individual nations to decide what should be done with disbursements received, just as they would be responsible for deciding on how to make participant fee payments. The money could be placed in the nation's general fund, earmarked for special projects, distributed to vessel owners, or used in some other way. Presumably the method of original fee payment would be considered by a nation in making its decision.

To clarify and illustrate the operation of a participant fee system, three detailed examples have been developed for the eastern Pacific yellowfin and skipjack fishery. The examples are limited to these two species for the sake of simplicity. They should not be taken to imply that no participant fees would be collected for other species harvested in this fishery such as bigeye and bluefin tuna. The first example given in the upper portion of Table 5 illustrates how the situation would look if a participant fee system were instituted in a fishery structured along the lines observed in the 1970–74 period. This first example, in which RAN catches are not related in any direct way to adjacent resources, enables comparison of the situation existing during the 1970–74 period with that which would apply if allocations were adopted as illustrated in the next two examples. These two examples are intended to cover the logical range of possibilities if it is assumed that RAN allocations are to be based on their adjacency to the resources as indicated by catches in their 200-mile zones. The second example in Table 5 illustrates how things would look if each RAN fleet developed to the point where its yellowfin and skipjack catch in the entire eastern Pacific exactly equaled the 1970–74 average catch made by the combined international fleet in its own 200-mile zone. This would be comparable to assuming that RANs receive high adjacency-based allocations for both species which they then fully utilize. While this example represents a logical but rather unrealistic extreme, it serves a valuable purpose in illustrating the potential seriousness of the situation facing non-RANs. The third example in Table 5 illustrates an opposite extreme in which RANs receive low yellowfin and skipjack allocations based on their own catches in their own coastal zones in the 1970–74 period and harvest only this amount of tuna. This is also an unrealistic situation, but together with the previous example, it does serve to define a range of possibilities for PAQ management with participant fees. The three examples can be referred to respectively as the 1970–74 catch distribution example, the high allocations example, and the low allocations example. In all examples RANs and non-RANs are grouped separately. Among the non-RANs, the United States is unique in that it would receive allocations and adjacency-based disbursements, but the amounts are so minute that they have no effect on its status as a non-RAN. The examples also assume that all RANs participate in the management system. Any RAN that did not participate would not have an allocation or share in participant fee disbursements. Also, non-RANs that fish beyond 200 miles but fail to cooperate in the program would not receive disbursements.

In the first section of the 1970–74 catch distribution example, catches of yellowfin and skipjack in the eastern Pacific are based on observed national catches. No catches are shown for RANs that were not participants in the fishery. Note that the total catch for each species equals the sum of the coastal zone and high-seas shares of the catch as given in the first part of Table 3. The next section shows participant fee payments made by each nation. Fees are assumed to be $40 per ton for yellowfin and $30 per ton for skipjack. These figures are believed to be close to the maximum that efficient harvesters could afford to pay under mid-1970s conditions of fleet size and resource availability in the eastern Pacific. At these fee levels, the payment for a full load of fish would roughly approximate the mid-1970s licensing cost in countries such as Ecuador and Peru.

Under the conditions outlined, 262,100 tons of catch would generate fees totaling a little less than $10 million—certainly not a particularly impressive sum, but probably close to the maximum that the system could generate under mid-1970s circumstances. From this total a deduction of 20 percent or a little less than $2 million is made

TABLE 5. EXAMPLES ILLUSTRATING COLLECTION AND REDISTRIBUTION OF PARTICIPANT FEES UNDER (1) 1970–74 CATCH DISTRIBUTION; (2) HIGH ALLOCATIONS; (3) LOW ALLOCATIONS

EXAMPLE 1 – CATCHES DISTRIBUTED AMONG RAN AND NON-RAN NATIONS AS IN RECENT YEARS.

Disbursements of participant fees after deduction of funds for management and enforcement

Harvesting nations / RANs	EPO catch Yellowfin (1000 s.t.)	Skipjack	Total	Fee Yellowfin ($40/ton)	Skipjack ($30/ton)	Total payment	Coastal Yellowfin	Coastal Skipjack	Coastal Total	Harvest Yellowfin	Harvest Skipjack	Harvest Total	Grand total	Net RAN benefits
RANs														(thousands of dollars)
Colombia	–	–	–	–	–	–	$195.1	$116.3	$311.5	–	–	–	$311.5	$311.5 (+)
Costa Rica	0.5	0.2	0.7	$20.0	$5.2	$25.2	726.9	250.1	977.0	$6.9	$0.6	$7.5	984.6	959.3 (+)
Ecuador	8.6	6.9	15.5	342.4	208.2	550.6	521.4	430.4	951.7	118.9	23.8	142.6	1,094.4	543.8 (+)
El Salvador	–	–	–	–	–	–	115.1	27.8	142.9	–	–	–	142.9	142.9 (+)
France (Clipperton)	2.0	1.6	3.6	79.2	48.5	127.8	121.3	6.2	127.5	27.5	5.5	33.1	160.5	32.7 (+)
Guatemala	–	–	–	–	–	–	131.1	24.6	155.7	–	–	–	155.7	155.7 (+)
Mexico	14.5	3.5	18.0	579.4	106.3	685.6	1,152.3	341.0	1,493.3	201.2	12.1	213.3	1,706.6	1,020.9 (+)
Nicaragua	6.9	3.4	10.3	276.2	101.1	377.3	53.6	20.8	74.3	95.9	11.5	107.4	181.7	195.6 (–)
Panama	1.5	0.9	2.4	61.6	26.4	87.9	256.7	50.2	306.9	21.4	3.0	24.4	331.3	243.4 (+)
Peru	–	–	–	–	–	–	209.9	167.7	377.6	–	–	–	377.6	377.6 (+)
RAN total	34.0	16.5	50.5	$1,358.8	$495.7	$1,854.4	$3,483.3	$1,435.0	$4,918.3	$471.8	$56.5	$528.4	$5,446.7	$3,592.3 (+)
Non-RANs														Net non-RAN deficits
Bermuda	1.5	0.9	2.4	60.0	27.0	87.0	–	–	–	20.8	3.1	23.9	23.9	63.1 (–)
Canada	7.1	3.6	10.7	285.4	108.2	393.6	–	–	–	99.1	12.3	111.5	111.5	282.2 (–)
Japan	0.8	0.3	1.1	33.1	8.4	41.4	–	–	–	11.5	1.0	12.4	12.4	29.0 (–)
Netherlands Antilles	1.4	1.9	3.3	56.2	58.2	114.3	–	–	–	19.5	6.6	26.1	26.1	88.2 (–)
Spain	5.9	3.5	9.4	235.4	104.8	340.3	–	–	–	81.8	12.0	93.7	93.7	246.5 (–)
U.S.A.	141.6	43.0	184.6	5,665.2	1,290.2	6,955.4	0.2	0.2	0.4	1,967.2	147.2	2,114.4	2,114.8	4,840.6 (–)
Non-RAN total	158.4	53.2	211.6	$6,335.4	$1,596.7	$7,932.0	$0.2	$0.2	$0.4	$2,199.9	$182.2	$2,382.1	$2,382.5	$5,549.6 (–)
Grand total	192.4	69.7	262.1	$7,694.1	$2,092.4	$9,786.5	$3,483.5	$1,435.2	$4,918.7	$2,671.7	$238.7	$2,910.5	$7,829.2	$1,957.3 (–) Cost of management
Deduction for management and enforcement (20%)				1,538.8	418.5	1,957.3								
Available for disbursement				6,155.3	1,673.9	7,829.2								

EXAMPLE 2 – RAN CATCHES EQUIVALENT TO CATCHES BY THE INTERNATIONAL FLEET INSIDE OF 200 MILES.

Disbursements of participant fees after deduction of funds for management and enforcement

Harvesting nations / RANs	EPO catch Yellowfin (1000 s.t.)	Skipjack	Total	Fee Yellowfin ($40/ton)	Skipjack ($30/ton)	Total payment	Coastal Yellowfin	Coastal Skipjack	Coastal Total	Harvest Yellowfin	Harvest Skipjack	Harvest Total	Grand total	Net RAN benefits
Colombia	6.1	4.8	10.9	$243.9	$145.4	$389.3	$195.1	$116.3	$311.5	$84.7	$16.6	$101.3	$412.7	$23.4 (+)
Costa Rica	22.7	10.4	33.1	908.7	312.6	1,221.3	726.9	250.1	977.0	315.5	35.7	351.2	1,328.2	106.9 (+)
Ecuador	16.3	17.9	34.2	651.7	538.0	1,189.6	521.4	430.4	951.7	226.3	61.4	287.7	1,239.4	49.7 (+)
El Salvador	3.6	1.2	4.8	143.9	34.7	178.6	115.1	27.8	142.9	50.0	4.0	53.9	196.8	18.2 (+)
France (Clipperton)	3.8	0.3	4.0	151.6	7.7	159.3	121.3	6.2	127.5	52.6	0.9	53.5	181.0	21.7 (+)
Guatemala	4.1	1.0	5.1	163.9	30.8	194.6	131.1	24.6	155.7	56.9	3.5	60.4	216.1	21.5 (+)
Mexico	36.0	14.2	50.2	1,440.3	426.2	1,866.5	1,152.3	341.0	1,493.3	500.2	48.6	548.8	2,042.0	175.5 (+)
Nicaragua	1.7	0.9	2.6	66.9	26.0	92.9	53.6	20.8	74.3	23.2	3.0	26.2	100.5	7.6 (+)
Panama	8.0	2.1	10.1	320.8	62.8	383.6	256.7	50.2	306.9	111.4	7.2	118.6	425.5	41.9 (+)
Peru	6.6	7.0	13.5	262.4	209.7	472.0	209.9	167.7	377.6	91.1	23.9	115.0	492.6	20.6 (+)
RAN total	108.8	59.8	168.6	$4,354.0	$1,793.8	$6,147.8	$3,483.3	$1,435.0	$4,918.3	$1,512.0	$204.6	$1,716.6	$6,634.9	$487.1 (+)

Net non-RAN deficits (thousands of dollars) — continued (Non-RAN rows, headers on preceding page)

Harvesting nations	Yellowfin	Skipjack	Total	Yellowfin ($40 per ton)	Skipjack ($30 per ton)	Total payment	Coastal Yellowfin	Skipjack	Total	Harvesting Yellowfin	Skipjack	Total	Grand total	Net non-RAN deficits
Bermuda	0.8	0.2	1.0	31.7	5.0	36.8	–	–	–	11.0	0.6	11.5	11.5	25.2 (–)
Canada	3.8	0.7	4.4	150.6	20.2	170.9	–	–	–	52.4	2.3	54.7	54.7	116.2 (–)
Japan	0.4	0.1	0.5	17.4	1.6	18.9	–	–	–	6.1	0.2	6.3	6.3	12.6 (–)
Netherlands Antilles	0.7	0.4	1.1	29.7	10.9	40.6	–	–	–	10.4	1.2	11.7	11.7	28.9 (–)
Spain	3.1	0.7	3.8	124.2	19.6	143.8	–	–	–	43.0	2.2	45.3	45.3	98.6 (–)
U.S.A.	74.7	8.0	82.7	2,986.4	241.3	3,227.7	0.2	0.2	0.4	1,036.9	27.5	1,064.4	1,064.4	2,162.8 (–)
Non-RAN total	83.5	10.0	93.5	3,340.1	298.6	3,638.7	0.2	0.2	0.4	1,159.8	34.1	1,193.9	1,194.3	2,444.4 (–)
Grand total	192.4	69.7	262.1	7,694.1	2,092.4	9,786.5	3,483.5	1,435.2	4,918.7	2,671.8	238.7	2,910.5	7,829.2	1,957.3 (–) Cost of management
Deduction for management and enforcement (20%)				1,538.8	418.5	1,957.3								
Available for disbursement				6,155.3	1,673.9	7,829.2								

EXAMPLE 3 – RAN CATCHES EQUIVALENT TO RECENT MAXIMUM CATCHES BY THEIR OWN FLEETS WITHIN 200 MILES OF THEIR OWN COASTS.

Disbursements of participant fees after deduction of funds for management and enforcement.

Harvesting nations	Eastern Pacific Ocean catches (thousands of short tons)			Participant fee payments (thousands of dollars)			To coastal nations on the basis of resource adjacency (thousands of dollars)			To harvesting nations on the basis of resource utilization (thousands of dollars)			Grand total	Net RAN benefits / Net non-RAN deficits (thousands of dollars)
	Yellowfin	Skipjack	Total	Yellowfin ($40 per ton)	Skipjack ($30 per ton)	Total payment	Yellowfin	Skipjack	Total	Yellowfin	Skipjack	Total		
RANs														
Colombia	–	–	–	–	–	–	195.1	116.3	311.5	–	–	–	311.5	311.5 (+)
Costa Rica	0.8	0.1	1.0	32.7	4.2	36.9	726.9	250.1	977.0	11.4	0.5	11.8	988.9	951.9 (+)
Ecuador	9.0	13.3	22.3	360.8	399.7	760.5	521.4	430.4	951.7	125.3	45.6	170.9	1,122.6	362.1 (+)
El Salvador	–	–	–	–	–	–	115.1	27.8	142.9	–	–	–	142.9	142.9 (+)
France (Clipperton)	0.0	–	0.0	1.6	–	1.6	121.3	6.2	127.5	0.6	–	0.6	128.0	126.4 (+)
Guatemala	–	–	–	–	–	–	131.1	24.6	155.7	–	–	–	155.7	155.7 (+)
Mexico	8.0	3.2	11.2	320.5	95.7	416.2	1,152.3	341.0	1,493.2	111.3	10.9	122.2	1,615.4	1,199.3 (+)
Nicaragua	–	–	–	–	–	–	53.6	20.8	74.3	–	–	–	74.3	74.3 (+)
Panama	1.3	0.7	2.0	53.0	20.8	73.8	256.7	50.2	306.9	18.4	2.4	20.8	327.7	253.9 (+)
Peru	1.6	1.9	3.5	62.2	58.4	120.6	209.9	167.7	377.6	21.6	6.7	28.3	405.9	285.3 (+)
RAN total	20.8	19.3	40.1	830.8	578.9	1,409.7	3,483.3	1,435.0	4,918.7	288.5	66.0	354.5	5,272.8	3,863.2 (+)
Non-RANs														
Bermuda	1.6	0.9	2.5	65.0	25.6	90.6	–	–	–	22.6	2.9	25.5	25.5	65.1 (–)
Canada	7.7	3.4	11.1	309.2	102.5	411.8	–	–	–	107.4	11.7	119.1	119.1	292.7 (–)
Japan	0.9	0.3	1.2	35.8	7.9	43.8	–	–	–	12.4	0.9	13.3	13.3	30.4 (–)
Netherlands Antilles	1.5	1.8	3.4	60.8	55.1	116.0	–	–	–	21.1	6.3	27.4	27.4	88.6 (–)
Spain	6.4	3.3	9.7	255.1	99.4	354.4	–	–	–	88.6	11.3	99.9	99.9	254.5 (–)
U.S.A.	153.4	40.8	194.2	6,137.3	1,222.9	7,360.3	0.2	0.2	0.4	2,131.2	139.5	2,270.7	2,271.1	5,089.2 (–)
Non-RAN total	171.6	50.4	222.0	6,863.3	1,513.5	8,376.8	0.2	0.2	0.4	2,383.3	172.7	2,555.9	2,556.4	5,820.5 (–)
Grand total	192.4	69.7	262.1	7,694.1	2,092.4	9,786.5	3,483.5	1,435.2	4,918.7	2,671.8	238.7	2,910.5	7,829.2	1,957.3 (–) Cost of management
Deduction for management and enforcement (20%)				1,538.8	418.5	1,957.3								
Available for disbursement				6,155.3	1,673.9	7,829.2								

NOTE: Sums of individual entries may not agree exactly with totals shown because of rounding.

to cover all management, surveillance, and catch monitoring costs. Comparing this amount with the fiscal 1974–75 budget of the IATTC ($790 thousand) and considering the substantial expansion of responsibility involved in implementing a PAQ management system, this management deduction certainly does not seem excessive. In these examples it comes to about $7.50 per ton of tuna harvested or roughly 1.0 percent of what the 1977 dockside value of this catch would have been.

The distribution of the $7.83 million remaining after deduction of management costs is shown in the remaining sections. Separate disbursements for yellowfin and skipjack are necessary because the participant fees are different for the two species, and they occur in different relative numbers in the coastal waters of various nations. Of the $7.83 million, $4.92 million is disbursed to RANs on the basis of resource adjacency, using the coastal zone catch proportions given in the first part of Table 3. In other words, adjacency-based disbursements made to RANs represent participant fees collected on fish caught within the 200-mile zone less a 20 percent deduction for management and enforcement. Non-RANs do not share in this stage of the disbursement process, and the amount disbursed is independent of allocation sizes or what is done with the remainder of the fees. The $4.92 million distributed in this manner represents 56.6 percent and 85.7 percent of the disbursable funds accruing from yellowfin and skipjack, respectively.

There remains $2.91 million representing the high-seas portion of the eastern Pacific catch. It is assumed that this is distributed among all harvesting nations, RANs and non-RANs alike, in proportion to their catches no matter where taken. The last column shows the net results of participant fee payments and disbursements. All RANs receive more in fee disbursements than they pay in. In fact, adjacency disbursements alone exceed payments for all RANs except Panama. (Panama's catches slightly exceed the average total catches made in its coastal zone in this example.) All non-RANs, of course, pay in more than they receive back. The United States, with by far the largest fleet, makes the largest payments, and despite the fact that it receives the greatest disbursements of any country, its net contribution (net deficit) is the greatest. Note that the difference between total non-RAN deficits and total RAN benefits exactly equals the deduction made for management, as it must.

In the second or high allocations example of Table 5, it is assumed that the tonnage of each RAN fleet changes by an amount that enables it to take catches of each species exactly equal to the average catches taken in its coastal zone in recent years by the entire international fleet. This means that fleets and catches increase in all RANs except Panama, whose catches drop slightly. The non-RAN share of the catch must go down because the total RAN catch increases substantially under the assumption of high and fully utilized allocations, while the overall eastern Pacific catch is assumed fixed. The non-RAN catch is distributed among nations according to the catch proportions in the current conditions example. A comparison of the expected catches in the first two examples shows that the combined RAN catch would more than triple (from 50,000 tons to nearly 170,000 tons), while the non-RAN share would be more than cut in half (from over 210,000 tons to less than 95,000 tons). United States losses alone would exceed 100,000 tons, representing nearly 40 percent of the combined yellowfin and skipjack catch in the eastern Pacific. Costa Rica and Mexico would be the big gainers, followed by Ecuador.

The remaining columns in the high allocations example are derived in the same fashion as corresponding columns in the current conditions example. Generally speaking, as compared to the first example which is based on the 1970–74 catch distribution, RAN participant fee payments are increased while non-RAN payments are down. Withdrawals for management and adjacency-based disbursements are naturally unchanged. Disbursements to harvesting nations based on high-seas catches show increases for RANs (except Panama) and decreases for non-RANs. Overall, the excess of disbursements received over fees paid declines for all RANs (except Panama), and net deficits of non-RANs are reduced. Payments by the United States, although reduced by over 50 percent, remain the highest of any nation; disbursements received by the United States also drop sharply and are lower than those received by Mexico, Costa Rica, and Ecuador. The net effect is that the United States deficit is reduced to 45 percent of its former value, although it is still much greater than that of any other non-RAN.

In the third or low allocations example of Table 5, RAN yellowfin and skipjack catches are assumed to equal the maximum annual catches made by their own fleets in their own coastal zones during the 1970–74 period (see second part of Table 3). With the exception of Ecuador, catches of all RAN participants are reduced from their average 1970–74 level in example 1. Ecuador's catch goes up because her fleets had exceptionally good fishing within her own 200-mile zone in certain years (on yellowfin in 1972 and on skipjack in 1971; see Table 2). Overall the total RAN catch is reduced by more than 10,000 tons; the non-RAN catch increases by the same amount and is distributed among nations in the same proportions as in the other two examples. Comparison of catches in the two allocation examples emphasizes the enormous differences between the high allocations and the low allocations approaches. RAN yellowfin and skipjack catches under high allocations are, respectively, 5.3 and 3.1 times larger than under low allocations.

The sections of the third example pertaining to collection and disbursement of participant fees were calculated just as in the first two examples, with differences for individual nations depending in an obvious fashion on differences in their catch levels. The United States deficit is maximized in this example, reflecting the dominant position that it would hold in the fishery.

One of the most interesting things about these three examples is the relatively small amount of money involved. The maximum total disbursement received by any nation in any example is only a little over $2 million (United States in examples 1 and 3, Mexico in example 2), and the maximum excess of disbursements received over fees paid is less than $1.2 million (Mexico, example 3). These figures underscore the fact that no nation would reap a vast bonanza of wealth from the tuna industry through a participant fee system. Furthermore, because of limits imposed by product values and harvesting costs, the same conclusion would apply to any kind of national licensing system or to any system based on transferring national allocations of the property right type. These examples, however, are based on the assumption of extreme competition within a very large fleet for a limited fixed resource. If fleets could be reduced sharply, fewer vessels could take the same catch fishing on a more efficient basis. Average annual catch per capacity ton and annual profit per capacity ton would both increase, and substantially higher fees (say double or more) might eventually become possible. Of course, any such fee increases would be a matter for negotiation.

Thus far in discussing participant fees it has been assumed that any allocations would be nontransferable guarantees of access to shares of the catch. If instead allocations were transferable property rights, modifications of the participant fee system would be necessary. If a nation could not utilize its full allocation, it could still derive benefits from the unutilized portion by transferring all or part of it to another nation in exchange for some agreed-upon payment. Clearly such benefits in the form of transfer payments should be in lieu of adjacency-based participant fee disbursements. But, except in the case of high allocations, RANs would derive benefits through participant fees based on coastal zone catches in excess of allocations. They would also receive disbursements based on catches of any unallocated species. A RAN should also retain the option of receiving participant fee disbursements for any unutilized part of its allocation that it chooses not to transfer. Perhaps the simplest way to handle this situation would be for transferred allocations to come under the participant fee system only to the extent of covering management costs. Catches made under such transferred allocations would be subject only to reduced participant fees covering management costs. Then when adjacency-based disbursements are made to RANs, the recent average catches by the international fleet in the 200-mile zones would be reduced by the amount of the transferred allocations in making all calculations. Finally, catches made under transferred allocations would not be considered in the last stage of disbursement when funds attributable to the high-seas portion of the catch are distributed among harvesting nations.

If differences among alternatives considered in this section seem relatively trivial in terms of the amounts of money involved in collection and disbursement of participant fees, bear in mind that the differences in terms of distribution of catches among nations and among individual harvesters are profound. These differences lie at the very heart of the international conflict over tuna management policy, and their resolution has life and death implications for national tuna industries as well as individual harvesters. The real importance of the participant fee concept is that it provides the key to establishment of a rational PAQ management system.

Other Aspects of PAQ Management

In the three preceding sections certain fundamental elements of a PAQ management system have been discussed in considerable detail. These elements are partial allocation of the overall catch, open access to all fishing grounds with international licensing, and a system of participant fees which provides both for management and for redistribution of resource-derived benefits among nations. These elements provide a basis for understanding, in general terms, how a PAQ management system could work. However, there are several other important aspects of PAQ management to be considered. These include: (1) procedures for determining when to close the open season; (2) skipjack management when yellowfin catches are controlled and partially allocated; (3) transitional changes in RAN and non-RAN fleets associated with adoption of PAQ management; and (4) control of fleet growth. Each of these topics requires thorough consideration, which entails getting into rather complicated details. Therefore, to expedite treatment of PAQ management in the main text, in-depth consideration of these subjects is presented in Appendixes I–IV, with major points being briefly summarized in the remainder of this section.

Under PAQ management the method used to determine when to close the yellowfin fishery raises interesting and extremely important questions that are treated in Appendix I. To achieve a predetermined overall quota, catches of vessels at sea would have to be monitored. When the total catch to date plus the estimated catch to be made under allocations and incidentally after closure equaled the overall quota, then the open season would be closed. The critical issue is whether or not RAN catches made prior to closure should be considered as part of their allocations. If they are to be counted and a RAN does not take its full allocation before closure, then the allocation would become a *de facto* catch limit, and the RAN would have no opportunity to compete for the share of the unallocated catch representing resources beyond 200 miles. On the other hand, if they are not to be counted and the entire RAN allocation is available after closure, then growth of RAN fleets would be greatly stimulated, and these fleets would compete very strongly with non-RAN fleets prior to closure. While the former situation might be considered inequitable to RANs, in the latter situation they would seem to have an unfair competitive advantage because their postclosure catches are guaranteed.

A possible means for resolving the closure dilemma would be to establish two-tier national fleets. Individual vessels could be designated as belonging to either a nation's "allocation" fleet or to its "open-season" fleet. All catches by a nation's allocation fleet, both before and after closure, would count against its allocation, and these vessels could operate only until the full allocation was taken, even if this occurred before closure. Catches by a nation's open-season fleet would not be considered as part of its allocation, but these vessels could not fish after closure even if part of the allocation remained uncaught. Under such a two-tier system, a RAN could develop its allocation fleet until it was capable of fully utilizing its allocation; thereafter, if it wished to compete for a share of the unallocated resource, it could develop an open-season fleet to compete on equal footing with non-RAN fleets.

The exact nature of the closure dilemma and its resolution will depend on the type of PAQ management system that is adopted. Factors such as the basis for RAN allocations, whether or not allocations are to be periodically updated, and their transferability are analyzed with respect to closure in Appendix I.

Under PAQ management, RAN claims to special treatment arising out of their adjacency to the tuna resources would be recognized in the case of yellowfin through both catch allocations and adjacency-based disbursements of participant fee receipts. In the case of skipjack, however, which at least initially would be unallocated, RANs would benefit only through participant fee disbursements. The differential treatment of the two species raises some important issues that are dealt with in Appendix II.

In the future it may become necessary to establish an overall catch quota for skipjack. Two key questions are: Should RAN skipjack allocations be established at that time? If so, at what level and according to what criteria? These questions should be resolved as fully as possible at the time PAQ management is initiated in order to preclude future conflicts that might jeopardize the management program. Non-RANs might argue that establishment of yellowfin allocations, especially if they are relatively large allocations, represents a major concession to RANs and that skipjack should never be allocated, even if eventually an overall skipjack quota must be established. On the other hand RANs could argue that if and when controls of skipjack become necessary, allocations related in some way to adjacency should be established just as in the case of yellowfin. However this matter is resolved, it should be kept in mind that because yellowfin and skipjack differ in their migratory behavior, distribution, and seasonal availability, there is no particular reason why the two species should be allocated on exactly the same basis.

Because skipjack are often heavily concentrated in Ecuadorian waters, and because her present fleet is heavily dependent on these skipjack resources, the problems that would face Ecuador under PAQ management are given special consideration in Appendix II. With open access and assuming no overall skipjack quota, a great deal of international effort would be concentrated on skipjack in Ecuadorian waters. This would result in greatly increased competition for Ecuador's fleet of smaller vessels which fish locally. Under these circumstances, Ecuador might refuse to participate in a PAQ management system. However, even if she did not participate initially, there are good reasons (e.g., development of her own fleet) why Ecuador might later decide to join. This possibility emphasizes the importance of leaving any PAQ management agreement open to later adherence by interested nations.

If a PAQ management system is adopted and yellowfin allocations are established, it is logical that RANs should be allowed to build up their fleets to harvest their allocations. Appendix III deals in detail with changes in fleet composition associated with transition to PAQ management. Because transitional changes and associated potential problems would be most significant if RAN allocations are high compared to present RAN catches, this situation is assumed throughout most of Appendix III, but this should not in any way be construed as an endorsement of high allocations.

There are two principal ways in which RAN fleets could develop —through intraregional flag changes from non-RANs and through new construction. In the eastern Pacific flag changes would have an advantage over new construction in that they would not increase the size of an already excessively large fleet. This could be quite important for non-RANs, whose share of the catch would be reduced by high allocations. Nevertheless, flag changes might be opposed in some non-RAN quarters. The shipbuilding industry and the fishermen themselves might oppose flag changes for fear of losing new business and their jobs, respectively. Also, non-RAN vessel owners not wishing to make flag changes themselves might be opposed to their competitors having the option. If these combined forces were sufficiently strong politically to prevent flag changes, all RAN fleet development would depend on new construction. In this case one might expect RAN fleet growth to be slow for lack of investment capital. However, with jurisdictional problems resolved and shares of the catch guaranteed to RANs under a PAQ management agreement, investor thinking could be optimistic, and new construction might proceed rather rapidly.

To explore quantitatively the fleet transition question in the most extreme situation, the high allocations example of Table 5 was extended. A number of situations were examined. With regard to allocations, two alternative were considered: high allocations for yellowfin only and high allocations for both yellowfin and skipjack. Also two options for RAN fleet growth were studied: all growth by flag changes from non-RANs and all growth by new construction. Finally, four alternatives for phasing in high RAN allocations were considered— immediate, fast, intermediate, and slow—all of which led to full allocations within ten years. If all combinations are considered, this leads to sixteen cases. For each case, transitional changes in RAN and non-RAN fleet sizes, catches, and annual catches per capacity ton were calculated over a ten-year period. These results are presented in detail in Appendix III. One important result is that the method of fleet growth has little effect on annual RAN catch rates, but makes a big difference for non-RANs. If only yellowfin are allocated, the annual non-RAN catch per capacity ton drops from 1.69 tons to 1.29 tons with flag changes, while in the case of new construction it falls to 1.00 tons. If both species are allocated, the comparable figures are 1.09 tons and .75 tons. The calculations also show large interim year differences among allocation phase-in alternatives.

The sharp declines in non-RAN catch rates suggested by the fleet transition example might appear to imply economic chaos. It is important to bear in mind, however, that the high allocation example was purposely constructed to illustrate the most extreme situation that could reasonably be postulated under PAQ management. In reality, there are a number of reasons why the impact of adopting PAQ management should be much less severe on non-RANs than this example suggests. These reasons are discussed in Appendix III and can be briefly listed as follows:

1. Allocations could be set below the high levels assumed in the example and in the case of skipjack might never be necessary.
2. RAN fleet development by flag changes from non-RANs could be encouraged over new construction.
3. Allocations could be phased in slowly.
4. Vessels that sink, retire, or depart to fish in other areas might not be fully replaced.
5. Postclosure fishing in other areas (e.g., the Atlantic Ocean) was not considered in the example.
6. Catches of other species such as bigeye, northern bluefin, bonito, albacore, and black skipjack were not considered in the example.

7. Eastern Pacific skipjack catches might increase as a result of open access and/or improved fishing techniques.
8. Unutilized portions of RAN allocations would remain available to non-RANs.
9. Incentives might be provided to encourage RANs not to utilize parts or all of their allocations, possibly using participant fees derived from catches beyond 200 miles.
10. Processors owning vessels might be able to offset possible fishing losses with profits in other phases of their operations.

An attempt to roughly evaluate the significance of these various mitigating factors concludes Appendix III. Simple and seemingly reasonable assumptions are made concerning several of the mitigating factors from the preceding list, and the effect of these assumptions on catch, effort, and annual catch per capacity ton is determined for both RANs and non-RANs. Very briefly, it is assumed that allocations are set below the maximal levels assumed in the high allocations example of Table 5, that RANs do not utilize their entire allocations, that RAN fleets grow in part through flag changes, that some vessels leaving non-RAN fleets are not replaced, and that non-RAN skipjack catches can be increased. Postclosure catches in the Atlantic and catches of other species in the eastern Pacific are also considered. The conclusion reached under these assumptions is that while adoption of a PAQ management system would result in substantial catch reductions for non-RANs, their annual catch per capacity ton might remain essentially unchanged.

Beyond the question of transitional changes in RAN and non-RAN fleets associated with adoption of PAQ management, there is the problem of controlling overall fleet size. This problem is considered in Appendix IV. Rational means must be developed for dealing with the serious excess capacity problems that exist in many of the world's tuna fisheries, including that in the eastern Pacific. It is presumed that some sort of system for licensing vessels will be involved. In developing a licensing system for control of fleet growth, a noncontradictory set of requirements and objectives should be explicitly spelled out. One possible approach might involve establishing a desired maximum fleet size, possibly below the size of the present international fleet. In seeking to attain the desired fleet size, development of RAN fleets to levels that would enable full utilization of their allocations should not be hindered, and no vessel presently participating in the fishery should be arbitrarily excluded. Also, apart from RAN fleet development, the same rules should govern all vessels. Within the framework of these general guidelines, a set of specific procedures should then be formulated for handling licensing transactions involving various types of flag changes, entrance of new vessels, and departures of vessels.

To illustrate the complexity of the controlling fleet size problem and the type of careful thinking necessary to resolve it logically, a detailed hypothetical system for implementing such control is developed. As usual, this illustrative example should not be construed as a proposal or recommendation. Very briefly, a license based on standardized carrying capacity would be issued to all vessels presently participating in the fishery. Also a pool of licenses would be established for issuance to anticipated new entrants into developing RAN fleets. Then thirty-three different types of licensing transactions are defined and grouped into ten categories. For each category, licensing procedures are defined and effects on existing RAN, non-RAN, and overall fleet sizes are noted. Effects on maximum potential fleet size are also indicated.

Appendix IV concludes with a discussion of several additional considerations and problems pertaining to control of fleet size. For example, transferable licenses could have considerable value, and several alternative means for executing transfers are considered from this point of view. Also considered are various possible means for reducing fleet size, including a policy of partial replacement of departing vessels. Another interesting problem is that different RANs may wish to pursue entirely different economic and social goals in developing their fleets. This could be accomplished in a manner compatible with control of fleet size under a split fleet concept such as that discussed in Appendix I. Finally, relationships between regional gear limitation and control of fleets on a globally coordinated basis are considered.

A successful PAQ management system must incorporate an effective enforcement capability to insure that all vessels of member nations comply with any regulations that are established. The basic enforcement functions for an international body administering a PAQ management system would be to maintain a surveillance system for monitoring vessel locations and to inspect landings. Means for achieving these goals were discussed in some detail in Chapter 4. Very briefly, a satellite system could be developed to monitor vessels of cooperating nations, and international inspectors could monitor landings. This could be done at a reasonable cost. Any observed violations would be reported to the flag nation involved for legal action. The international body would not itself impose penalties, and it would exercise great care in separating its research activities from its enforcement program. Member nations would be individually responsible for preventing intrusions into their 200-mile juridical zones by fleets of nonmember nations.

To conclude this rather lengthy discussion of PAQ management, it is noteworthy that numerous problems and goals have been identified. In this chapter and Appendixes I-IV, many alternatives are considered for solving these problems and achieving these goals—some explored in considerable depth and others simply mentioned. In part, the purpose is to emphasize that management problems are significant, complicated, and important to solve. It is hoped that the concepts and ideas considered can serve as a point of departure leading ultimately to realistic and viable solutions. While there is definitely no intent to advocate on behalf of any particular approach, some approaches seem more promising than others as is clear from the discussion.

9 Regional Coalitions

The IATTC program to manage yellowfin tuna in the eastern Pacific has been in effect since 1966. The objective of the program has been to produce the maximum yield from the resources on a sustained basis, and it has been successful to the extent that serious overfishing of the resources has been prevented. However, as noted in Chapters 3 and 7, the program is plagued by serious shortcomings which place it in jeopardy. In addition to the IATTC program, some nations in the eastern Pacific region that are not IATTC members have attempted to manage tuna fisheries in their own coastal zones by imposition of high license fees. The problems of achieving effective management under this unilateral national control approach have also been pointed out (Chap. 6).

Because neither of these approaches provides an adequate means for future tuna management, it is necessary to seek new approaches. One such approach involving establishment of a PAQ management system has already been described in considerable detail (Chap. 8 and Appendixes I–IV). It was developed with the dual objectives of providing recognition for RAN claims to special consideration based on their adjacency to the resources while at the same time allowing open access to the entire tuna resource for all participants in the fishery. Many concepts, problems, and possible innovations were considered including, among others, national catch allocations related to resource adjacency, international licensing, a system for collecting and distributing participant fees, and a system for controlling future fleet size.

While some version of a PAQ management system might operate successfully, other alternatives for future management must also be considered. Various types of regional coalitions among nations will be analyzed in this chapter, first coalitions involving only RANs and then coalitions including both RANs and non-RANs. For the purposes of this discussion, a zone of extended jurisdiction to 200 miles will be assumed. As usual the emphasis will be on the situation in the eastern Pacific, although much of what is said would apply equally to other areas. Within this region, recall that the ten coastal nations from Mexico south to Peru (including France because of her possession, Clipperton Island) are considered RANs, while the United States, Canada, and Chile are considered non-RANs.

RAN Coalitions with Exclusion of Non-RANs

It will be recalled that control of fishing within 200 miles by coastal states acting individually was concluded to be impractical in several important respects (see Chap. 6). For example, in the absence of international cooperation there would be no management program, and the large non-RAN fleet would undoubtedly fish very intensively beyond 200 miles. Besides providing competition for RANs, such uncontrolled fishing might damage stock productivity, especially in the case of yellowfin. Another objectionable feature for RANs was that they would not have access to one another's 200-mile zones if all zones were exclusive or if high license fees were required. Because the annual and seasonal availability of tuna is highly variable, such exclusion would adversely affect development of RAN fisheries. On the surface it might appear that RANs could do little to overcome the problems of no management, non-RAN competition, and possible overfishing; but certainly the problem of exclusion from one another's zones could be attacked on a cooperative basis. This fact alone provides substantial motivation for RANs to enter into a reciprocal agreement or coalition, and this motivation will be stronger in the future than at present. RAN fleets, generally speaking, are small compared to the average resources available within their respective coastal zones, but as they grow, improved access to the resources will be increasingly important.

At the maximum an eastern Pacific RAN coalition would consist of the ten RANs, while at the minimum it could consist of any two of these nations. The willingness of a nation to join a proposed coalition would, of course, depend on the terms of the coalition and what the nation perceived as its gains and losses under those terms. There would be no sense in forming a coalition unless members derive special privileges of some sort from participating. Because the most obvious privilege would be the right to fish freely within the 200-mile zones of other coalition members, it is of interest to speculate as to which types of nations might be included in such a coalition.

It is fairly obvious that nations with relatively large fleets such as Mexico and Ecuador would stand to gain from a coalition with one another. Each of these nations plans further development of its fleets based in part on the fact that tuna are abundant within their own 200-mile zones in many years. There are years, however, when tuna are less abundant (see Table 2), and in such years it would definitely be advantageous to be able to seek fish off the coasts of other nations. Thus, nations with large fleets and substantial resources have both something to gain (flexibility for their fleets) and something to contribute (access to substantial resources in certain years). Assuming a coalition including Mexico and Ecuador, what would be their attitude toward the inclusion of other types of nations, for example those with large resources and relatively small fleets such as Costa Rica? Or those with fleets that are large in comparison to rather small resources such as Panama? It seems logical that Mexico and Ecuador would like to see Costa Rica included in a coalition because this would substantially increase their access to the resource base, and in certain years such as 1971 and 1974 this could be of vital importance. Costa Rica's attitude, however, is less predictable. If she plans to develop her fleet substantially beyond its 1977 level, then the future benefits of being able to fish off Mexico and Ecuador would provide a strong incentive to join; but if she has no such plans, she might prefer instead to license vessels (both RANs and non-RANs) to fish within her 200-mile zone. Panama's situation is quite different. She might logically want to participate in the coalition because access to Mexican, Ecuadorian, and Costa Rican waters would be valuable to the Panamanian fleet. But while gaining much, Panama would be contributing little in terms of access to resources. In fact, the main effect of Panamanian participation on other members would be to increase the level of competition facing their own fleets. However, considering such factors as Panama's strategic central location and the desirability of unified Latin American action, exclusion of Panama from a RAN coalition seems unlikely from a political point of view. Nations with both small fleets and small resources remain to be considered. Their participation or nonparticipation would not be much of an issue unless they had definite plans to enlarge their fleets substantially. Then they would fall in essentially the same class as Panama and would probably want to be included.

Under the circumstances just described, it is impossible to say just how inclusive an eastern Pacific RAN coalition might be. It is clear though that certain barriers, arising out of differing national circumstances, do exist to the formation of an all-inclusive coalition based only on reciprocal fishing rights. In the PAQ system previously described such differences in national circumstances would be taken into account by establishing catch allocations and a system for collecting and redistributing participant fees. Could such an approach be adopted by a RAN coalition? There is no apparent reason why allocations should be created for RAN coalition members in the absence of participation by non-RANs (and perhaps some RANs as well) in management and conservation of the resources. However, coalition members could set up a participant fee system among themselves with collections and disbursements being based solely on catches made by members within their common 200-mile zone. Under such a system the objections of some nations to being included and the disadvantages associated with participation of other nations could both be at least partially overcome. Costa Rica, for example, could derive benefits from participating even without a large fleet of its own, while Panama would make participant fee payments in excess of the disbursements that it would receive. Hence, a participant fee system might result in more nations joining a RAN coalition. In any case, a coalition would alleviate the problem of RAN fleets being excluded from one another's waters to some extent, depending on which RANs were involved. Additionally, policing of coastal waters to prevent poaching by vessels of non-RANs and nonmember RANs, though still a difficult problem, could possibly be handled more efficiently by a coalition than by each nation maintaining an independent enforcement system.

With these benefits in mind, what changes might be expected in the scenario that was developed for the case in which all RANs individually maintain exclusive zones with no coalitions? Interestingly, the scenario does not change greatly. Non-RAN vessels would fish very heavily beyond 200 miles and new techniques, especially for taking skipjack, might be developed. This outside fishery would provide competition for all RANs, but because coalition members would have access to one another's waters, the competitive effect on them could be considerably reduced as compared to nonmembers. Initially the outside catch would increase, but this increase would be offset by the decline of the inside catch owing to exclusion of non-RAN fleets. RAN fleets, especially those of coalition members, would likely grow, and the inside catch would increase. Overfishing of yellowfin would be possible, resulting in reduced catches and economic suffering for all. The extent of the reductions and resulting hardships would depend on the extent and method of RAN fleet expansion as well as on the spawner-recruit (S-R) relationship for yellowfin.

To more quantitatively evaluate changes that might result if a coalition consisting of all RANs were to exclude non-RANs entirely from the common 200-mile zone, a computerized model for simulating the eastern Pacific yellowfin fishery was utilized. In the model the fishery is divided into three areas—an inshore area corresponding roughly to the 200-mile zone and two offshore areas, one corresponding approximately to the CYRA beyond 200 miles and the other representing the fishery west of the CYRA. The innermost area is occupied by an inshore yellowfin stock from which there is partial emigration to the central area. The outermost area is populated by an offshore yellowfin stock that is also partially emigratory to the central area. Spawning in the innermost or outermost areas produces recruits in these areas. Spawning in the central area where the inshore and offshore stocks are intermingled provides recruits to all three areas. Growth, natural mortality, and fishing mortality are incorporated into the model using parameter estimates developed from historical data. Fishing mortality rates are age specific and are defined separately for each area and quarter of year (i.e., for each time-area stratum). The model can be run with a S-R relationship or in a constant recruitment mode (i.e., no S-R relationship takes effect at the levels of fishing effort under consideration). Three situations were explored using the simulation model:

1. Management policies and distribution of fishing effort continue as at present. Some fleet growth beyond the present level is projected in the model.
2. RANs exclude non-RANs from their common 200-mile zone and there is no management. Non-RANs fish unrestricted beyond 200 miles, while RANs fish unrestricted in the 200-mile zone. Fleet growth is as projected in situation 1.
3. Similar to situation 2, but it is assumed that there is a further doubling in RAN fleet size over a ten-year period.

Each of these three situations was simulated both with and without a yellowfin S-R relationship. Yellowfin catches during the tenth year are given for each of the simulations in Table 6. No information on the skipjack fishery is incorporated in the simulation model, and the catch estimates given in Table 6 for this species are based on general knowledge concerning their availability and distribution. Comparison of these results with the examples of Chapter 8 on PAQ management are somewhat difficult because different assumptions are made concerning RAN fleet growth, the S-R relationship, distribution of fishing effort, and so on.

Comparing results for the first two situations in Table 6 suggests that if RANs exclude non-RANs from the 200-mile zone and there is constant recruitment, then inside yellowfin catches would fall by nearly 75,000 tons while outside catches would increase by 62,000 tons for a net loss of about 13,000 tons. With a S-R relationship, outside catches would increase by only 35,000 tons, and the total yellowfin catch would fall by 39,000 tons. (There is no change in inside area catches because under the assumptions of the model the RAN fleet does not generate enough effort for the S-R relationship to come into play.) Interestingly, the simulation suggests that yellowfin catches would decline for both RANs and non-RANs. Exclusion from the 200-mile zone would have a severe impact on non-RAN skipjack catches. Because most skipjack catches are taken in the coastal zone, it is estimated that the non-RAN skipjack catch would fall by nearly 38,000 tons while the RAN catch would remain unchanged.

If we combine the data for both species and assume a S-R relationship for yellowfin, inside catches fall by 115,000 tons while outside catches are up by only 38,000 tons for a net loss of 77,000 tons. This large net loss is shared by both RANs and non-RANs, with RAN catches dropping by 9,000 tons and non-RAN catches by 68,000 tons. In the case of constant recruitment, the total net loss and the non-RAN losses would be less, 50,000 tons and 41,000 tons respectively. RAN losses would remain at 9,000 tons.

In the third simulated situation, RAN fleets are assumed to gradually double in size over a ten-year period. Such fleet growth does not seem unlikely and it is of interest that even with this much growth the model indicates only a modest 16,000 ton increase in RAN yellowfin catches as compared to the present situation. Non-RAN yellowfin catches would fall to even lower levels than in the case of no RAN fleet growth. The overall catch for both species combined would still be below present levels, especially if there is a S-R relationship.

All of these modeling results are in general agreement with earlier

TABLE 6

SIMULATED ANNUAL CATCHES BY RAN AND NON-RAN FLEETS (Thousands of short tons)

	Skipjack	Yellowfin		Yellowfin and skipjack combined	
		Without S-R relationship	With S-R relationship	Without S-R relationship for yellowfin	With S-R relationship for yellowfin
Situation 1: Fishery managed as in 1975					
Catch inside 200 miles	63.0	116.1	116.1	179.1	179.1
Catch outside 200 miles	7.0	105.5	105.5	112.5	112.5
RAN catch	22.5	50.3	50.3	72.8	72.8
Non-RAN catch	47.5	171.3	171.3	218.8	218.8
Total catch	70.0	221.6	221.6	291.6	291.6
Situation 2: Non-RANs excluded from 200-mile zone, no management					
RAN catch inside 200 miles	22.5	41.6	41.6	64.1	64.1
Non-RAN catch outside 200 miles	10.0	167.6	140.6	177.6	150.6
Total catch	32.5	209.2	182.2	241.7	214.7
Situation 3: Same as situation 2 but RAN fleets double in size over a 10-year period					
RAN catch inside 200 miles	45.0	66.7	66.7	111.7	111.7
Non-RAN catch outside 200 miles	10.0	158.6	129.9	168.6	139.9
Total catch	55.0	225.3	196.6	280.3	251.6

qualitative scenarios describing what might happen if management were to fail in the eastern Pacific. Nevertheless, it must be remembered that the simulation model is only a rough approximation of reality. More specifically, if non-RANs are excluded from the 200-mile zone, the model may underestimate the competitive effect of the intense fishery beyond 200 miles on the inside RAN fishery. Also the estimated fishing efficiencies used to describe fishing mortality in the model may be biased because they were estimated assuming no restrictions on gear movement between the inside and outside areas. This is not a reasonable assumption if non-RANs are excluded from the 200-mile zone. Finally, no provision is made for RAN vessels to fish outside their common exclusive zone, although at times they might do so.

The simulation results, even as limited as they are, indicate that formation of an exclusionary RAN coalition falls far short of being an acceptable approach to the problems of tuna management in the eastern Pacific. The plight of both non-RANs and nonmember RANs would be as serious as in the case of each RAN controlling its own exclusive zone. The circumstances of member RANs, while probably somewhat improved, could hardly be considered ideal. The catch allocation problem would remain unresolved, and there would be no management of the resource as a whole. Without international cooperation, no fleet limitation program would be possible. Finally, 90 to 95 percent of all tuna harvested in the eastern Pacific is marketed in the United States. If excluded from RAN 200-mile zones, it is possible that she might prohibit importation of tuna taken by RAN members. Even if not fully effective, such sanctions could be a severe blow to RANs. For all of these reasons, if regional coalitions are to provide a workable solution, further modifications in their organization and operation must be considered.

RAN Coalitions with Licensing of Non-RANs

In discussing control to 200 miles by individual coastal states, the possibility of a RAN licensing non-RANs and other RANs to fish in its coastal zone was considered. Two basic motivations for establishing licensing systems were noted. An obvious motivation was to produce income. This would be especially important to RANs with fleets that were small compared to resources available in their waters, and it would be the only consideration for RANs without fleets. On the assumption that fees would not be so exorbitant as to be exclusionary, another motivation for RANs with fleets to issue licenses would be to provide access to one another's waters. In the case of a RAN coalition issuing licenses, this latter motivation would no longer be present, at least with respect to coalition members, since they all would presumably have open access to one another's waters. The former motivation, however, remains important. The present combined RAN fleet is not large enough to fully utilize resources available within their collective 200-mile zone. Hence, it would be natural for a RAN coalition to license non-RANs (and conceivably also nonmember RANs) to fish for the available surplus.

A RAN coalition could have further motivations for issuing licenses not applicable to the case of licensing by individual RANs. For example, because of seasonal and annual variations in abundance and availability of tuna, an individual nation could not possibly issue or withdraw licenses rapidly enough to effectively control the level of competition faced by its own fleet within its 200-mile zone. A RAN coalition, however, might eventually want to structure a licensing system so as to restrict or control competition within their common 200-mile zone. If the coalition included most or all of the RANs, especially those with larger resources, control of competitive effort through licensing could probably be carried out quite effectively.

A RAN coalition might have another important goal in establishing a licensing system. If all license purchasers, as a condition of issuance, were required to participate in a program for management and conservation of the resource as a whole, then conceivably the problem of uncontrolled fishing beyond 200 miles could be resolved, the total catch regulated, and possible damage to the stocks by overfishing averted. These highly desirable potential results would provide a strong incentive for a RAN coalition to develop a licensing system keyed to management of the resource. This concept will be pursued later in this section, but first the situation in which a coalition issues licenses to generate income and perhaps control competition will be considered on the assumption that there is no overall resource management program.

Suppose a RAN coalition decided to institute a licensing program based, say, on registered net tonnage of nonmember nation vessels. One important decision would involve setting the fee level. Presumably the members would be seeking that fee level which, while being consistent with other goals, would maximize revenues to the coalition. Obviously it would make no sense to set fees so high that no one could afford to pay them and still hope to fish profitably. By the same token, a coalition would not want fees significantly below the level that any purchaser would be willing to pay. These bounds define a range of fee levels over which some, but not all, vessels would be attracted to purchase licenses, and to maximize income one must consider both the fee level and the number of licenses likely to be sold at each level. A further consideration, at least eventually, might well be to set fees high enough so as to limit competition from nonmember license purchasers, especially if growth of fleets of coalition members is to be encouraged.

As noted in the preceding section, a RAN coalition might wish to institute a system of participant fees based on catches made within the common 200-mile zone so that members without developed fleets could still derive benefits from the resources. If so, nonmembers could be required to pay these participant fees in place of or in addition to licensing fees, subject to the one requirement that members should have some financial advantage over nonmembers. If licenses were free and participant fees the same for both members and nonmembers, then the advantage of members would be limited to receiving all fee disbursements. It is more likely that nonmember vessels would be required to pay license fees in addition to participant fees in order to insure an advantage for member nations at the individual vessel level and to limit the number of nonmember vessels. Alternatively, nonmember participant fees could be made higher than those paid by members.

Under any of these licensing systems (i.e., nonmembers pay license fees only, pay a combination of license fees and participant fees, or pay only participant fees with free licenses), if a RAN coalition wishes to control competition from nonmember vessels fishing in its common 200-mile zone, it will have to decide how many licenses to issue. While high license or participant fees could reduce competition, a more desirable alternative might be to limit either the number of licenses issued or the catch made by licensed vessels. This approach would have the advantage of being more exact in its application, and the number of licenses issued would not be subject to fluctuations owing to changes in fishing costs and/or product value (unless fees are so high that some licenses remain unissued).

In addition to collecting fees, coalition members would have to agree on how to distribute license and/or participant fees and would also need to develop an enforcement capability to prevent violation of their 200-mile zone. After deduction of enforcement costs, disbursements could reasonably be based on catches made by all participants in the fishery within the 200-mile zones of member nations. Enforcement could be accomplished more efficiently if centralized rather than left in the hands of individual members. Enforcement costs would depend on the type of system adopted. A simple licensing system with neither member or nonmember participant fees would be least expensive to enforce, while systems involving participant fees would be more costly because of the need to quantitatively monitor when and where catches are made. Coalition members, either individually or collectively, would also have to develop a capability to apprehend violators.

A RAN coalition could issue licenses to vessels of any non-RAN without making distinctions among them. On the other hand, licenses might be issued according to some sort of elegibility criteria. For example, in the eastern Pacific, non-RANs could be grouped according to geographical location and/or historical participation in the fishery. Then vessels from certain national groups could be allowed to purchase licenses and others excluded. Or the national groups could be placed in a hierarchy with licenses being issued to vessels from the top group first, then, if any remained, to vessels from the next group, and so on. In the eastern Pacific, the highest category might logically be regional non-RANs which have historically participated in the fishery (i.e., the United States, Canada, and Chile). The next group might be other historical non-RAN participants, with all remaining non-RANs constituting the final group. Another approach would be to license only vessels presently participating in the fishery. Also, a coalition might refuse to license vessels of nonmember RANs, thus providing a strong inducement for nonmembers to join.

Attitudes of different RANs toward licensing by a coalition could be influenced by the size of a nation's fleet, the magnitude of its coastal resources, and its plans with regard to fleet growth. Nations such as Mexico and Ecuador with relatively large and growing fleets and substantial resources might logically be concerned primarily with controlling competition faced by their fleets. Their interest in generating revenue would probably be secondary. If so, they would naturally want any licensing system structured so as to reflect these priorities. A nation like Costa Rica with a relatively small fleet and large resources might take a position similar to that of the large-fleet, large-resource RANs if it had near-term fleet expansion plans, but revenues would also be a major concern, particularly if rapid fleet expansion was not contemplated. Hence, Costa Rica might favor a licensing scheme that would allow more competition, but would also generate more revenue. Because of her important resources, other RANs would most likely want Costa Rica included in the coalition, so her views could carry substantial weight.

RANs with limited coastal zone resources would fall into two categories: those which have or intend to develop fleets that are large in relation to their coastal zone resources and those without such fleets and no plans to develop them. The latter would seem to have only a very minor stake in the structure of a licensing system and would most likely accept whatever other coalition members might decide upon. Nations of the former type, best exemplified by Panama, might be more concerned with controlling competition than with generating revenue.

Considering the differing interests of the various RANs and still assuming that there is no overall management program, how might a coalition licensing system evolve? Initially and for as long as RAN fleets are small compared to total coastal zone resources, the objective of maximizing revenues would certainly be an important one to a RAN coalition. Even during this period, however, RANs would probably want some competitive protection from the large non-RAN fleet in the eastern Pacific. Hence, rather high and at least partially exclusionary fees (either license fees or nonmember participant fees) might be expected, and, perhaps also, limits on the number of licenses or on the magnitude of nonmember catches.

As RAN fleets develop, it is reasonable to expect that the objective of limiting competition would become more and more dominant over the revenue objective, especially since this would likely be the main objective of the larger and more dominant coalition members. A licensing system would probably evolve toward being ever more exclusionary and eventually might be abandoned altogether. Hence, licensing could ultimately lead to essentially the same undesirable result as the simple exclusion of nonmembers discussed in the previous section. Although the scenario might take a few years longer to unfold, non-RANs would fish heavily beyond 200 miles, providing competition for fleets of RAN coalition members. Without overall management of the fishery a likely result would be overfishing on yellowfin, with economic chaos and suffering plaguing RANs and non-RANs alike.

Coalition members might avoid some of these problems by strictly limiting the size of their own fleets, but a far better solution would be to eliminate the root cause of all the problems, the lack of any kind of cooperation for management of the resource as a whole. As suggested earlier, achievement of such management could provide an alternate motivation for establishing a licensing system. Policies for management could be established by a RAN coalition that would apply to the whole resource. As a condition of being issued a license to fish within the 200-mile zone, nonmembers would have to agree to abide by all regulations that the coalition might promulgate.

Obviously, for such coalition management policies to be effective, a substantial part of the nonmember fleet would have to cooperate. This means that there should be a strong incentive for vessels to seek licenses and hence participate in the management program. To provide such an incentive a coalition would probably want to set nonmember license or participant fees at a relatively low level. Thus, the objective of achieving management of the resource through licensing seems to some extent incompatible with the objective of maximizing license or participant fee revenues to the coalition. Similarly, the goals of management and stringent control of competition also appear at least partially incompatible.

A RAN management program could take various forms. Perhaps the simplest approach would be for a coalition to issue licenses for appropriate fees to nonmember vessels allowing them to fish within the common 200-mile zone until closure of the fishery. After closure, licensed vessels would cease fishing operations in the eastern Pacific both inside and beyond 200 miles, but they would not be prevented from transferring to other ocean areas. In determining the time of closure, outside catches of nonparticipating vessels both before and after closure would have to be considered as well any catches made legally after closure under any special allowances that the coalition might grant.

More elaborate management systems would also be possible, and to the extent that they more adequately resolved the catch distribution problem, they would indeed be probable. For example, a combination of participant fees and adjacency-related allocations for coalition members might be adopted to insure RANs access to shares of the catch, to equalize benefits among RANs, and to meet management costs. The same participant fee would probably be collected for all catches of a given species regardless of whether they come from inside or outside 200 miles. This would eliminate any incentive for reporting inside catches as being outside and vice-versa. Redistribution of fees could then be a three-step process similar to that described in Chapter 8 in connection with PAQ management. First, management costs would be covered; then fees derived from coastal zone catches (including licensing fees if any) would be distributed to RANs on the basis of adjacency; and finally fees derived from catches beyond 200 miles would be distributed to all cooperating participants, RANs and non-RAN licensees alike, on the basis of their catches. Alternatively, if participant fees were disbursed only to coalition members, non-RANs could logically argue that fees on outside catches should only cover management costs. Differential fees would raise monitoring problems, but though troublesome, these might not be insoluble. A coalition would also have to address many of the questions raised in discussing PAQ management (e.g., transferability or nontransferability of allocations, method of closure, possible control of fleet size, multispecies management problems, etc.). Much of this discussion also applies here and need not be repeated.

Assuming that a RAN coalition were to adopt a licensing system with its primary objective being to secure management over the entire resource, how likely is it that such a system could succeed? One obvious problem is that nonmembers would have no real voice in management decision making. Even though a RAN coalition would probably solicit suggestions from nonmembers, decision making authority would ultimately rest with the RAN members alone. Nonmembers, especially non-RANs with large fleets, might find this objectionable. For one thing, nonmembers might argue that if RANs have all the management authority then they should bear all the management costs, including enforcement costs which for an effective system would be substantial. This position would imply no participant fees for management or else such fees for coalition members only. On the other hand, coalition members could argue that licensed nonmember vessels derive immediate and long-term benefits from effective enforcement, which prevents poaching by nonlicensed vessels and provides for effective management, and that therefore they should share management costs. In any case, the coalition would have the power to impose management fees. But whether nonmembers would be willing to cooperate in licensing and management under such arrangements would remain to be resolved.

A second problem associated with concentration of all authority in the hands of RAN coalition members is that as their fleets grow, the objective of limiting competition is likely to increase in importance, quite possibly to the detriment of the management objective. Strong pressures could develop within a coalition to move in the direction of more exclusionary licensing through higher fees, limits on the number of nonmember licenses, or limits on nonmember catches. Faced with this future possibility, nonmembers before agreeing to cooperate in a management program might seek assurances that exclusionary modifications would not be made in the licensing system. The extent to which a coalition would be willing to give such assurances, however, and their real value are open questions. Perhaps the most fruitful approach would be for coalition members to establish allocations for themselves and then voluntarily key fleet growth to these allocations.

These problems make it clear that a RAN licensing and management system would have to overcome major obstacles in order to secure continuing nonmember participation on a scale large enough to insure success, for without such participation the familiar scenario of competition from the fishery beyond 200 miles, possible resource depletion, and economic suffering can simply be reiterated. To secure cooperation, meaningful consideration would have to be given to the major problems facing nonmembers, especially those with large investments in the fishery; concessions would have to be granted and assurances given to ameliorate those problems. It seems possible, but not necessarily probable, that a tenable accommodation might be worked out between a RAN coalition and nonmembers. Chances of success would certainly be enhanced if nonmember governments, especially governments of non-RANs with large fleets, required or at least encouraged their flag vessels to cooperate in a coalition management regime. But without a policy making voice, nonmember governments might not be willing to do so.

Coalitions of RANs and Non-RANs with Allocations

In most regional tuna fisheries, RANs hoping to benefit from the migratory stocks that come and go in their 200 mile coastal zones through development of their own fishing industries will face competition from large, highly efficient, and already well-entrenched non-RAN fleets. Certainly this is the case in the eastern Pacific where the United States purse-seine fleet dominates the fishery. A RAN coalition would be ill-advised to simply exclude such fleets from their coastal waters because effective management of the resource as a whole would be impossible. Issuance of licenses to nonmember vessels willing to cooperate in a management program could conceivably resolve this problem, but it is questionable whether non-RANs would be willing to cooperate with coalition management in which it had no policy voice.

This leads logically to consideration of regional coalitions that include non-RAN members.

In the eastern Pacific the United States with its dominant fleet is the most obvious non-RAN candidate for inclusion in a regional coalition, but two other nations, Canada and Chile, could also logically be included. These two nations border on the Pacific, but, more importantly, Canada is a significant participant in the fishery, while tuna resources (mainly skipjack) of potential commercial significance occur in Chilean waters and probably belong to the same stocks that are exploited further to the north. Also, Chile has historically participated in the fishery. A closed coalition consisting of all Latin American RANs, France, the United States, Canada, and Chile would include all thirteen coastal states of the eastern Pacific region.

The RANs would have a strong rationale for seeking to include the United States in a regional coalition. Assuming that she did participate as an equal member with a policy voice, United States cooperation in management would be assured. She would share in covering the costs of the research, monitoring, and enforcement programs so essential to successful management. More importantly, with the United States participating as a government, the entire United States fleet would be required to comply with coalition regulations. This would greatly reduce the problems of competition and possible overfishing resulting from fishing beyond 200 miles by noncooperating vessels before and after closure.

In structuring a regional management system, the needs of both RAN and non-RAN coalition members would have to be considered. The principle requirement for RANs would probably be recognition of their claim to special treatment arising from their adjacency to the resource. The principal requirement for non-RANs would probably be to secure long-term access to the resources wherever they become abundant and available either within or beyond the 200-mile zone. All members would benefit from maintaining the total harvest at or near its maximum level through careful management.

A program aimed at meeting these requirements could vary in its particulars, and there is no need to explore all possibilities here. However, certain general observations can be made. Some form of adjacency-related allocations would at least partially meet the requirement of RAN members for special treatment by insuring them the opportunity to develop their own fishing industries if they chose to do so. Hence, throughout the remainder of this section, such allocations will be assumed. Some possible alternatives not involving resource allocation will be considered in the following section. An allocation system could be accompanied by participant fees for insuring more equitable distribution of benefits among RAN members. Participant fees could also cover management and enforcement costs. Finally, a coalition of RANs and non-RANs might want to adopt measures for control of fleet size. Such a coalition management system would closely resemble the PAQ management system discussed in Chapter 8. Most of the concepts introduced in that discussion concerning problems and options involved in establishing a management program would apply with equal validity to management by a closed regional coalition. There is one important difference, however. Whereas participation in the PAQ system would be unrestricted, only RANs and regional non-RANs would participate in the type of coalition presently under consideration. The obvious questions are: Would regional non-RANs, especially the United States, be willing to participate in such a closed regional coalition? If so, could such a coalition succeed in sustaining viable management over the resources? To answer these key questions, three important problem areas must be considered.

The first problem, and an especially disturbing one for the United States, would be the possibility of regional non-RANs gradually being squeezed out of the fishery. It is logical that RAN allocations would encourage RAN fleet growth, and it is possible that as their fleets grew some RANs might demand increasingly larger shares of the catch. Under these circumstances, and especially if excluded from other regions, the United States would have no real choice but to withdraw from the coalition and compete as effectively as possible by fishing beyond 200 miles. Why then should the United States participate in a regional coalition under circumstances that would favor RAN fleet growth? Would it not be better to meet the challenge of increasing RAN competition head-on and immediately in the hope of preventing or at least limiting further expansion of RAN fleets? The drawback to this line of thinking, of course, is that it leads to the same conclusion that was noted in discussing control of coastal zones by either individual RANs or a RAN coalition (i.e., no management, uncontrolled competition between the inside and outside fisheries, possible overfishing, economic suffering, etc.). Clearly any assurances that could be given to the United States that she would not be gradually squeezed out of the fishery would mitigate in favor of her joining a coalition rather than going it alone. Interestingly, adjacency-based allocations could lead to one type of assurance if RANs agreed that such allocations would constitute permanent upper limits on the catches that would be guaranteed available to RANs.

The problem of gradual non-RAN exclusion has been considered from the United States point of view. What of the remaining eastern Pacific non-RANs, Canada and Chile? Canada's position would probably parallel the United States position because her fleets could also be subject to gradual exclusion by RANs. Chile, however, might not share this concern, at least as long as she remains a nonparticipant in the fishery itself. But if she were to develop a significant fleet, her attitude might move toward that of the United States and Canada.

The second problem affects all coalition participants and involves the tacit assumption that with the inclusion of the United States and Canada, competition from vessels of nonmember nations fishing beyond 200 hundred miles would be insignificant. While vessels from beyond the eastern Pacific region constitute only a relatively small part of the present fleet, and while some of these might fish elsewhere if a closed regional coalition were formed, nevertheless the fact that substantial resources are available beyond 200 miles cannot be ignored. In the period 1970–74 yellowfin and skipjack catches beyond 200 miles averaged well over 90,000 tons annually (Table 2). Over a period of time resources of this magnitude could attract substantial effort to compete with regional coalition members. Some nonregional vessels already fishing in the eastern Pacific might remain there. Also, additional nonregional vessels from both presently participating and newly entering nations might join the fishery. The USSR, for example, has a number of longliners and purse seiners under construction in Poland with delivery scheduled between 1978 and 1981. It is possible that a segment of this fleet will fish in the eastern Pacific, perhaps in a mothership operation. But ironically, the greatest potential source of noncooperating vessels could prove to be the fleets of the members themselves, especially the United States fleet. If a coalition failed to control the size of its combined fleet, economic conditions in the regulated fishery could become so severe that vessel owners might seek nonregional flags of convenience under which they could fish beyond 200 miles on an unregulated, year-around basis.

If competition from nonmembers became significant, a regional coalition would face an unpleasant choice. If overall production is to be maintained at close to a maximum level, catches taken in the uncontrolled outside fishery would have to be estimated and deducted from the amount remaining available to fleets of coalition members operating under the constraint of a conservation program. Alternatively, if an effort was made to maintain coalition catches in the face of nonmember competition beyond 200 miles, it is likely that overfishing would take its toll (at least in the sense of reduced yield per unit of recruitment) and that everyone's catches would begin to fall. It is difficult to see how coalition members could accept either of these unpleasant choices, and unless remedies were found it seems unlikely that a closed regional coalition could remain viable in the face of a significant unmanaged fishery beyond 200 miles.

To preclude development of an unmanaged fishery beyond 200 miles, members of a closed regional coalition might attempt to negotiate some sort of agreement with potential competitors, perhaps backed up by the threat or actual employment of economic sanctions of various kinds. Another possibility is that formation of closed regional coalitions could come to be accepted internationally as an appropriate approach to tuna management. In this case, the principle of regional control over regional resources would be internationally recognized, and nonregional nations as a matter of policy would not allow their flag vessels to fish except under the auspices of a region's management program.

Suppose a closed regional coalition could somehow avoid the problem of gradual exclusion of member non-RANs and, additionally, could succeed in excluding or at least sharply restricting competition from nonmembers fishing beyond 200 miles. A third major problem, involving the matter of precedent, would still remain. If eastern Pacific RANs and non-RANs formed a closed regional coalition and excluded all other nations from fishing within the 200-mile coastal zone, nations in other regions could certainly be expected to follow suit and exclude members of the eastern Pacific coalition. While Latin American RANs might not find the immediate effects of this too disturbing, the implications of being excluded from other regions would be very serious for the United States and Canada, many of whose seiners are in some years heavily dependent on catches made in the eastern Atlantic, especially after closure in the CYRA. Virtually all these catches are made within 200 miles of various African RANs. For example, in 1974 these catches amounted to roughly 26,000 short tons of yellowfin and skipjack. United States flag vessels have also recently begun to fish in the western Pacific. If United States and Canadian vessels are prevented from fishing in coastal zones of other regions, problems associated with excessive fleet carrying capacity in the eastern Pacific could be aggravated, even considering that exclusion of nonregional flag vessels would be an offsetting factor.

Because of the precedent problem and the problem of potential nonmember competition beyond 200 miles, the nations of the eastern Pacific might well prefer to allow nonregional flag vessels to fish in the eastern Pacific on a managed basis. This could be accomplished through licensing or by inclusion of nonregional nations in a regional management coalition. The licensing alternative requires little elaboration since much of what was said in the preceding section concerning licensing by a RAN coalition would also apply to licensing by a closed regional coalition. Of the licensing goals considered earlier, that of insuring cooperation in a resource management program would be of most significance. In addition there would be the important new goal: to establish a precedent for access through licensing to resources of other regions.

Enlarging a regional management coalition to include nonregional nations provides an interesting alternative to licensing vessels from such nations. The first group of nations to receive consideration might logically be nonregional historical participants in the fishery. In the eastern Pacific these would be Japan, Spain, Bermuda, the Netherlands Antilles, Senegal, Venezuela, and, most recently, New Zealand and the Congo. (These extraregional historical participants might possibly be accorded something less than full membership status.) Inclusion of historical participants could go a long way toward resolving the problem of competition from nonmember nations, especially if all choose to participate in the coalition, because any unregulated fishing by newly entering nonmember nations in the eastern Pacific would have to develop from scratch (except that which might conceivably arise through transfer of presently active vessels to flags of convenience). The precedent problem would not be so neatly resolved, however. While the United States and Canada are established as historical participants in the eastern Atlantic fishery, other eastern Pacific nations are not so established and would be permanently excluded from an Atlantic coalition limited to regional nations plus historical participants. This could have serious consequences for a nation like Mexico whose developing fleet may soon wish to fish in the Atlantic off Africa. There could also be disagreement over how long a historical nonregional participant should be allowed to continue in a fishery. For example, would it be fair for a nation with only a year or two of experience in a given fishery to participate indefinitely while nations which had not participated had no rights at all? Hence, while inclusion of historical participants might tend to resolve or diminish certain problems, other problems are raised. This leads naturally to consideration of further broadening a regional management coalition.

After inclusion of RANs, regional non-RANs, and historical participants in a regional coalition, there seems to be no apparent reason for singling out any other particular group of potential entrants. Thus, the broadest type of coalition to consider is one in which any nation with an interest in the fishery is free to join. This would include all RANs whether they fish or not. All non-RAN participants in the fishery could also be included regardless of whether they come from the region in question or are extraregional, and regardless of whether they are historical participants or new entrants. Under an open-to-all approach the precedent problem would be resolved, but the coalition would still have to deal with problems relating to distribution of catches, control of fleet size, enforcement and surveillance, and so forth.

As in the case of other types of coalitions, the actual organization of an open-to-all coalition could take many alternative forms and could incorporate many of the concepts discussed elsewhere. Presumably there would be some system of adjacency-related allocations to protect the interests of RAN members. There might also be participant fees to support management and to provide for equalization of benefits among RANs. The unallocated share of the catch could be available to all, and to prevent excessive competition a system for limiting or even reducing fleet carrying capacity could be adopted. While such an approach would require considerable cooperation among nations, it is clear that any nation with an interest in maintaining an effectively functioning open-to-all coalition in any given region would have a considerable incentive to insure that its flag vessels cooperated in regional management efforts elsewhere.

In considering various possible coalitions of RANs and non-RANs, a very interesting and important conclusion has been reached: An open-to-all regional management coalition could operate in a manner that would make it indistinguishable from an analogous PAQ management system. Thus, in pursuing the subject of regional management by coalitions of nations, essentially the same end point has been reached as was reached in Chapter 8 on PAQ management when the perspective was strictly international.

Coalitions of RANs and Non-RANs without Allocations

In the preceeding section it was noted that a successful coalition of RANs and non-RANs would require that basic needs of both classes of nations be adequately met. In particular, RANs would insist upon some form of special consideration arising out of their adjacency to the resource. It was then assumed that to fulfill this requirement, at least in part, adjacency-related allocations would be established. Another approach has been suggested, however, in the case of the eastern Pacific fishery. Instead of national allocations, an overall quota could be established and all coalition members would fish on equal footing until closure. After closure each RAN would have the special privilege of continuing to fish without restriction for the rest of the year, but only within either its own or the common 200-mile zone. Thus, closure would occur when the catch to date plus estimated catches to be made by RANs after closure equaled the quota. A coalition operating on this basis could be limited to regional coastal states (i.e., in the eastern Pacific the ten RANs plus the United States, Canada, and Chile), or it could include other historical participants, or it could be open to all nations.

First, suppose that each RAN was allowed to continue fishing only within its own 200-mile zone after closure. This would obviously be the most advantageous arrangement from a non-RAN point of view because each RAN would be restricted to the minimum possible area after closure. This limitation would restrict postclosure catches, and smaller postclosure catches imply larger preclosure catches and a later closure date. Hence, limiting of the area that could be fished by each RAN after closure would also limit the period of time during which RANs could fish without competition from non-RANs.

Non-RAN catches under such an arrangement would almost certainly remain comparatively high, and they might naturally favor the approach. A RAN with a large fleet relative to its resources might also be in support, but Panama is presently the only such RAN in the eastern Pacific. RANs such as Ecuador and Mexico with partially developed fleets and comparatively large resources in their coastal zones could probably continue their fishing operations at roughly their current levels. But their fleets already can fish after closure under the present IATTC regime (either under special allowances or for skipjack) without being restricted to their own coastal zones, so the trade-off of special allowances for restricted postclosure fishing privileges would hardly constitute a significant gain. Hence, Mexican and Ecuadorian support for such a proposal seems doubtful. Other RANs would gain little or nothing. In fact, the restricted area approach could actually curtail growth of their tuna industries, so they most likely would be in opposition. Considering the interests of all nations, the advantages to some are obvious and substantial, and the same can be said for the disadvantages to others. Clearly, the overall chances of forming such a coalition are slight.

It is also worth noting that management decision making would be difficult if each RAN was restricted to its own coastal zone after closure. In order to set a closure date, management would have to estimate postclosure catches in the coastal zone of each RAN. In order to do this, the fishing capability of the various RANs would have to be determined, a relatively simple task. But serious problems would arise in forecasting the distribution and availability of fish in national zones after closure. Because of seasonal and year-to-year variation in these factors, the task of setting a closure date would be fraught with uncertainty.

As an alternative to restricting postclosure RAN fishing to each RAN's own 200-mile zone, suppose instead that all RANs could continue to fish throughout their common 200-mile zone. This would make determination of a closure date simpler for management, but of greater importance are the implications for the RANs and non-RANs. While the restricted area approach would sharply limit RAN catches and quite possibly curtail RAN fleet development, with postclosure RAN fishing in the common 200-mile zone this would no longer be the case. Significant resources are often available somewhere within this zone throughout the year. Unrestricted access to these resources wherever they might be available would stimulate RAN fleet growth. Assuming that their fleets did develop, RAN catch shares would increase, closure would occur earlier, and non-RAN catch shares would gradually decline. As this process continued, non-RANs eventually might abrogate the agreement and, in the absence of management, concentrate great competitive effort beyond 200 miles to the possible detriment of the resource itself. Thus, an agreement allowing postclosure RAN fishing throughout the common 200-mile zone, while initially beneficial to RANs and detrimental to non-RANs, could lead ultimately to failure of management and economic disruption for all participants in the fishery. This approach might prove just as unacceptable as re-

TABLE 7

AVERAGE MONTHLY CATCH RATES FOR YELLOWFIN AND SKIPJACK IN NATIONAL 200-MILE ZONES AND BEYOND 200 MILES BY PURSE SEINERS OVER 400 TONS, 1970–74

Average Yellowfin Catch per Day of Effort (short tons)

Within 200 miles of	January	February	March	April	May	June	July	August	September	October	November	December	Annual
Chile	-	-	-	-	-	-	-	-	-	-	-	-	-
Colombia	5.34	6.59	8.03	10.96	5.22	-	-	-	-	-	-	-	7.34
Costa Rica	12.55	14.73	15.48	13.72	12.55	5.99	3.62	-	-	5.50	4.74	-	12.66
Ecuador	9.12	7.23	8.62	6.88	-	3.75	-	-	-	-	3.31	6.12	7.85
El Salvador	-	-	8.08	15.33	-	-	-	-	-	-	-	-	13.03
France (Clipperton)	15.54	7.27	14.41	-	-	-	-	-	-	-	11.75	-	13.36
Guatemala	-	-	16.93	17.82	-	-	-	-	-	-	-	-	13.59
Mexico	8.45	10.41	11.60	11.70	7.35	14.68	7.14	-	3.29	-	8.44	2.91	11.12
Nicaragua	-	-	-	-	-	-	-	-	-	-	-	-	5.79
Panama	4.63	5.42	7.91	11.86	7.48	1.36	-	-	-	-	-	-	7.93
Peru	4.59	6.32	5.82	-	-	-	-	-	-	-	-	-	6.87
U.S.A.	-	-	-	-	-	-	-	-	-	-	-	-	-
Within 200 miles	10.59	10.28	12.72	12.54	9.86	6.82	6.26	4.87	4.77	5.03	5.24	7.46	10.72
Beyond 200 miles	15.68	13.75	17.23	13.73	11.32	6.17	-	13.63	-	12.61	16.74	10.68	14.88
Entire CYRA	12.70	11.67	14.53	13.25	10.32	6.55	8.15	5.67	6.06	7.71	9.53	8.95	12.08
West of CYRA	-	-	-	14.13	12.70	13.39	12.81	14.05	12.68	8.82	9.58	24.80	11.58
Entire eastern Pacific	12.70	11.67	14.53	13.29	11.05	10.10	11.37	13.28	12.32	8.79	9.42	10.39	11.88

Average Skipjack Catch per Day of Effort (short tons)

Within 200 miles of	January	February	March	April	May	June	July	August	September	October	November	December	Annual
Chile	–	–	–	–	5.91	–	–	–	–	–	–	–	–
Colombia	3.91	0.91	3.28	3.45	5.50	–	1.85	2.73	–	1.42	3.55	0.57	3.49
Costa Rica	0.98	0.80	2.16	2.54	5.50	3.04	4.57	1.15	4.51	4.58	2.99	0.24	3.52
Ecuador	6.00	5.56	3.06	2.60	–	11.58	7.10	4.05	0.60	3.04	4.92	4.52	5.08
El Salvador	–	–	5.33	2.97	–	–	–	–	–	–	–	–	1.93
France (Clipperton)	0.46	0.94	1.25	–	–	–	–	–	–	–	0.35	–	1.01
Guatemala	–	–	2.19	1.98	–	–	–	–	–	–	–	–	1.42
Mexico	0.49	0.25	1.72	3.18	1.70	1.85	1.86	3.18	1.77	0.18	0.56	0.35	1.43
Nicaragua	–	–	–	–	–	–	–	–	–	–	–	–	3.96
Panama	0.10	0.49	1.11	4.15	4.13	2.53	–	–	–	2.18	–	–	2.66
Peru	11.56	4.74	5.76	–	–	–	–	–	–	–	–	–	6.81
U.S.A.	–	–	–	–	–	–	–	–	–	–	–	–	–
Within 200 miles	4.06	3.28	2.65	3.34	5.50	5.43	3.10	3.01	2.16	2.86	4.60	3.06	3.78
Beyond 200 miles	0.85	1.38	2.76	4.15	4.08	3.22	–	1.75	2.18	0.85	0.67	1.50	2.50
Entire CYRA	2.90	3.20	2.83	3.72	5.36	5.42	2.86	2.98	2.09	2.30	3.80	2.76	3.48
West of CYRA	–	–	–	0.57	2.41	0.88	0.50	0.98	0.81	0.57	1.25	3.49	0.88
Entire eastern Pacific	2.90	3.20	2.83	3.71	5.04	3.58	1.53	1.39	1.12	1.09	2.95	2.80	2.88

NOTE: In the case of yellowfin, effort directed primarily at skipjack after closure is excluded from consideration.
A dash indicates insufficient data (i.e., less than 3 years in which there were more than 10 days of effort expended).

stricting postclosure RAN fishing rights to each nation's own coastal waters.

In considering the previous alternatives, no distinctions were made among the various non-RANs. They could be divided, however, into coastal non-RANs within a region under consideration and non-RANs from outside the region. Such a distinction could be meaningful if a geographical region is defined as extending beyond the area actually occupied by the harvestable resource. For example, for purposes of yellowfin and skipjack management, the eastern Pacific region could be defined to include waters off the United States, Canada, and Chile even though recent catches in these waters are negligible. These three nations could then be considered coastal non-RANs. As coastal nations, these non-RANs could argue that they should legally be entitled to the same rights and privileges as regional RANs which are also coastal nations. Note that in Chapter 8 on PAQ management this position was tacitly recognized in discussing adjacency-based allocations and participant fee systems because allocations and disbursements, even though minuscule, were provided for the United States.

Within the present context, what are the implications of common treatment for all regional coastal states, both RANs and non-RANs alike? Suppose postclosure fishing privileges were restricted to each nation's own 200-mile zone. Except for possible legalistic significance, such treatment for the United States and Canada would have little practical impact because yellowfin and skipjack resources are virtually nonexistent off their coasts. But in the case of Chile, the matter could be more than academic. In the future, meaningful postclosure catches might be made in Chilean waters, and she would certainly want the right to fish in her waters after closure should she enter the fishery. If Chilean catches did become significant, her status would change from that of a coastal non-RAN to that of a RAN. In any event, granting coastal non-RANs postclosure fishing privileges in their own zones would not change any of the conclusions previously reached.

On the other hand, if postclosure fishing rights within a common 200-mile zone applied to coastal non-RANs as well as to RANs, the scenario would be drastically altered. Unrestricted fishing by the United States and Canada within the common 200-mile zone after closure would make control of catches utterly impossible. Closure would occur very early in the year because of the large catches that would be taken in the common 200-mile zone. In fact, there might not even be an open season. Non-RANs from noncoastal states would be excluded from the fishery even if they abided with the management program, but it is doubtful that they would. Considering these consequences, it is difficult to imagine international agreement to such a system.

If postclosure fishing privileges in coastal waters are to be considered as a possible alternative to adjacency-related allocations, it is logical that RANs will want to evaluate the potential value of such privileges. If fish, generally speaking, are unavailable in a given area during the time that a special fishing privilege applies there, then the privilege is not worth much. Of course the converse is also true. Hence, in making an evaluation it is logical to consider past fishing success in time-area strata of interest. For the eastern Pacific, Table 7 gives the catch of yellowfin and skipjack per day of fishing effort by purse seiners over 400 tons in carrying capacity by months and by national 200-mile zones. The data shown are averages for the five-year period, 1970–74. Because annual figures are highly variable, no data based on ten or fewer days of effort are included in the averages, and no average is given unless it is based on at least three or more such years.

Table 7 does not require extensive comment. However, the frequent absence of adequate data, especially after closure in the spring, is noteworthy. Also catch rates are generally higher beyond 200 miles than they are inside. Inspection of annual averages and averages over the entire eastern Pacific suggests that the international fleet is capable of maintaining a relatively high average catch rate by moving from place to place as availability of fish dictates. This points up the importance of maintaining access to the entire eastern Pacific tuna resource, especially for non-RANs. Also, the value to a RAN of being able to fish in its own 200-mile zone after closure is often limited owing to poor availability of resources or uncertain owing to lack of information.

10 Total Allocation
of the Resource

Catch distribution systems considered so far involve either partial allocation of the available resource to individual nations or no allocation at all. Control by individual RANs over exclusive 200-mile zones would fall into the no-allocations category as could control by regional coalitions. Under the present IATTC management system, roughly 25 percent of the yellowfin quota is taken under allocations, but these allocations are based on various economic factors rather than on resource adjacency. Partial allocation of the allowable catch on the basis of resource adjacency was discussed thoroughly in connection with PAQ management. Under this approach, national allocations could be set at any agreed-upon level with a reasonable upper bound being recent average catches within 200 miles of the various RANs, or about 57 percent of the yellowfin catch in the eastern Pacific for the 1970–74 period. Management by a regional coalition could also involve adjacency-based allocations.

Under any of these approaches, some part of the resource ranging from all of it down to the portion taken beyond 200 miles would be available to all participants in the fishery, RANs and non-RANs alike, on a competitive basis. While free competition encourages innovation and technological advance, it also injects an element of uncertainty into the picture in the sense that the share of the unallocated catch that a nation takes in one year may change in the next, possibly for reasons beyond its control such as competition from new entrants into the fishery. This suggests the possibility of allocating the entire catch among participants in the fishery. Certainly considerable precedent exists for this approach to resolving the catch distribution problem, and three well-known examples of total catch allocation at the international level will be examined briefly. These examples involve northern fur seals, world whale resources, and the assemblage of species taken in the north-

120

west Atlantic, and the insights they provide could be pertinent to the tuna situation.

The earliest international agreement for total allocation of a marine resource involves fur seals and goes back to 1911. Fur seals, which originally were harvested on their island rookeries in the Bering and Okhotsk seas, became the subject of an intensive and very wasteful pelagic fishery in the latter part of the nineteenth century, and the resource was seriously depleted. Under the original agreement between the United States, the USSR, Japan, and Great Britain (acting on behalf of Canada) seal harvesting was restricted to young males and took place exclusively on the rookeries. The harvest was shared among nations exercising sovereignty over the islands on which rookeries occurred and former pelagic sealing nations. Under management, stocks recovered and production increased. The original treaty was abrogated during World War II, but was renewed in 1957 when the four original participants ratified a convention creating the North Pacific Fur Seal Commission (NPFSC). At present, the United States and the USSR, the sovereign states responsible for managing and harvesting the resource, each retain 70 percent of the production from their own rookeries, with the remaining 30 percent being divided equally between Japan and Canada. Thus, total allocation of the resource is carried out on the basis of sovereignty over the resource, investment in management and utilization, and historical participation in the former pelagic fishery. In earlier years, willingness to refrain from pelagic sealing was also a factor, but this is no longer a real consideration because the economic viability of a pelagic fishery is doubtful under present conditions.

The International Whaling Commission (IWC) established in 1946 includes both whaling and nonwhaling nations. It was established to oversee the conservation and utilization of the world's whale resources. For a number of years the IWC assessed the condition of whale stocks and recommended an overall catch quota for baleen whales in the Antarctic which were the subject of international concern because of their rapid depletion, especially in the case of the blue whale. However, it had no power to insure adherence with its recommendations or to allocate catches. Instead, in 1962 the five nations actively whaling in the Antarctic agreed among themselves on the distribution of the overall IWC catch quota. The USSR claimed 20 percent of the catch based on the size of her recently enlarged whaling fleet. The remaining 80 percent was divided among Norway, Japan,

Great Britain, and the Netherlands, primarily on the basis of their recent historical catches. These allocations were transferable, and at the end of the original four-year agreement Japan had acquired all of the Netherlands and British shares and part of the Norwegian share. When the IWC agreement was renegotiated in 1966 (and annually thereafter), national allocations were established. They were based on fishing capability and were nontransferable. Originally, overall quotas and national allocations for the Antarctic were expressed in terms of the so-called "blue-whale unit," which was used to interrelate the relative oil yields of the several species sought. While agreement to national quotas represented an important step forward, adoption of the blue-whale unit proved to have unfortunate consequences from a conservation point of view. Under the formula defining the unit, the relative values of the different species were not equal. Hence, effort was directed primarily toward taking blue whales, the relatively most valuable species, until they had been depleted to the point of commercial extinction. As blue whale catches fell, effort shifted first to the finback whale and then to the sei whale, the next two most valuable species, with similar results. Faced with the failure of this approach, the IWC turned to species quotas, first in the North Pacific and then in the Antarctic, and these have proven to be much more effective in terms of conserving whale resources. Increasing world pressure to reduce quotas to levels that would not further deplete the stocks has also played a significant role. A limited amount of whaling takes place outside this IWC framework, but not yet enough to threaten stocks or jeopardize the degree of international cooperation that has been achieved. The IWC experience suggest the desirability of establishing quotas and allocating catches on a species-by-species basis.

The International Commission for Northwest Atlantic Fisheries (ICNAF) was formed in 1950 to manage and conserve the numerous commercially important species taken in the northwest Atlantic Ocean from Greenland to New England. Its 1975 membership of eighteen nations included all nations that fished in the northwest Atlantic. Haddock stocks, which had been seriously depleted, were the first to be protected under an ICNAF quota in 1970. The haddock quota and the yellowtail flounder quota which was added in 1971 were both overall quotas, or, in ICNAF parlance, global quotas. The first ICNAF national allocations were established for herring in 1972, and many other species were allocated from 1973 on. Originally national allocations were determined by distributing 40 percent of the total allowable catch on the basis of catches during the preceeding ten years, 40 percent on the basis of catches during the last three years, and 10 percent among the coastal states (note that the United States and Canada are RANs in this instance), with the final 10 percent being reserved for new entrants. This formula is no longer followed; each allocation is now negotiated separately among whichever nations are interested in harvesting the particular species in question. Total allowable catches are established for discrete stocks which may extend across several ICNAF management areas. When this is the case, national allocations are made for each area separately in order to distribute fishing effort. An interesting recent development has been ICNAF's recognition of the fact that catches of all species cannot be maximized simultaneously because of interspecies interactions such as predation and competition. To account for this, a second kind of national allocation has been established for all except coastal nations. These are "all-species" allocations set at less than 100 percent of the sum of a nation's species specific allocations. Each nation decides for itself how to distribute its catch among species. If a nation takes its full allocation of some species, it must take less than its full allocation of others in order to stay within its all-species allocation.

These examples—northern fur seals, whales, and the northwest Atlantic fisheries—suggest that the actual or imminently threatened depletion of a resource is one possible motivation for adopting a total allocation system. With total allocation, no matter what the basis, the role of every participating nation is clearly defined and there is a high degree of management control over the fishery. Such control would not normally be expected in a developing fishery on an underharvested species or stock.

Other factors can motivate adoption of a total allocation system. With total allocation, each nation's role is defined and uncertainties associated with extremely competitive fisheries are reduced. From certain points of view the assurance that a specified catch can be taken could be appealing. For example, suppose a RAN wishes to develop a fishing industry. In an intensely competitive fishery, the success of a new entrant may be far from certain, and a prudent investor might hesitate to risk his capital in such a venture. However, if the RAN were to receive an allocation, perhaps based on adjacency, guaranteeing access to some specified share of the total catch, the same prudent investor might be eager to participate because the guarantee would eliminate much of the uncertainty and reduce his risk. The same reasoning would apply to replacement of a non-RAN vessel given that the non-RAN had an allocated share of the resource, perhaps based on historical participation in the fishery. Hence, total allocation might appeal to some nations as a way to stimulate development of their fleets, at the same time appealing to others in terms of securing a continuing role as a participant in the fishery.

In totally allocating a resource, two basic questions must be answered. First, which nations will participate in the allocation? Second, what will the criteria for allocation be? In answering the first question, an obvious possibility would be to limit participation to regional nations and perhaps also historical participants in the fishery. Another possibility would be to leave the door open for future new entrants either by setting aside a portion of the catch for this purpose (as is done by ICNAF) or by providing for reallocation upon request to create a share for a prospective new entrant. Regardless of whether allocation is on a closed or an open basis, the rights of participating nations to divide the resource among themselves would have to be generally recognized internationally. If noncooperating nations were able to enter the fishery and take substantial catches, any total allocation scheme could founder. International recognition would be especially essential in the case of tuna allocation, because in many areas substantial catches are made beyond 200 miles where there would be no national authority even if extension of fisheries jurisdiction to that distance were universal. In this respect, tuna resources are more similar to whale resources than they are to the demersal and nearshore pelagic resources managed by ICNAF.

A number of criteria have either been actually used or else proposed as a basis for total allocation. Some of these such as resource adjacency, historical participation in the fishery, and fleet size were mentioned in discussing northern fur seals, whales, and ICNAF resources. Other criteria that could be considered include population size, gross national product, consumption of the catch from a particular fishery, investment in the fishery, and economic need. Some of these would be easy to quantify; others, for example economic need, would be quite difficult to measure. The final choice for a specific fishery would depend to some extent on the decision as to which nations are to participate in the allocation.

In the case of the eastern Pacific tuna fishery, IATTC member governments have spent considerable time since the inception of the management program discussing possible criteria for total allocation of the yellowfin resource. The most frequently discussed criterion has been adjacency to the resource. This has been expressed in terms of coastline length and the distance to the center of distribution of the resource or in terms of catches within 12 or 200 miles of coastal nations. The second most frequently discussed criterion has been historic participation in the fishery as measured by both long-term and short-term catch histories. The obvious reason for the emphasis on these two criteria is that they exemplify the differences between RANs with developing tuna industries and technologically developed non-RANs. Any allocation scheme based on the length of coastline adjacent to the resource or on catches in adjacent waters would work to insure shares of the resource for developing RAN fleets. On the other hand, any allocation scheme based on historic participation would insure that the non-RANs who developed the fishery would retain a substantial share of the catch.

Under total allocation, if RAN claims to shares of the resource were ultimately met by establishment of adjacency-related allocations, then non-RANs might well insist that they be allocated the remaining catch on the basis of historical participation. In this connection, it is interesting to note that in the mid-1960s the United States share of the total eastern Pacific yellowfin catch was just over 90 percent. At that time Latin American RANs expressed the philosophy that they were entitled to special consideration because of their adjacency to the resource. These claims were not generally recognized; instead, beginning in 1969, various special allowances were granted to different classes of vessels based on economic hardship. These special allowances have steadily increased, while the United States share of the yellowfin catch has gradually been reduced to its 1975 level of about 70 percent, a figure which includes substantial United States catches taken west of the CYRA. Had the United States negotiated a total allocation agreement in the mid-1960s involving some recognition of RAN adjacency claims, it is very probable that she could have secured a share of the catch for herself exceeding that which she could secure at present assuming that similar negotiations were undertaken. It is also possible that total allocation of the catch at that time might have served to discourage nonregional nations from entering the fishery.

An elaborate scheme for total allocation of the allowable yellowfin catch was discussed at a special meeting of the IATTC in 1972. Under this scheme the following criteria would be used in allocating portions of the catch:

1. Each nation's catch during the preceding three years
2. Total catch within the 12-mile zone of each coastal state
3. Total catch from 12 to 200 miles of each coastal state
4. For countries bordering on the CYRA, allocations of the catch between 12 and 200 miles based on population
5. For countries bordering on the eastern Pacific, allocations of the catch between 12 and 200 miles based on population
6. For countries bordering on the Pacific Ocean, allocations of the catch beyond 200 miles based on population
7. For all countries participating in the fishery, allocations of the catch beyond 200 miles based on population

Different weights could be assigned to each of these criteria so that allocations could be adjusted to their final values over a period of years. Under one such allocation scheme the United States during the first year would have been alloted 76 percent of the CYRA yellowfin catch, with 8 percent to Mexico, 7 percent to other IATTC members, and 9 percent to nonmember participants in the fishery. Thereafter, annual adjustments over a nine-year period would have resulted in long-term allotments as follows: United States, 47 percent; Mexico, 30 percent; other IATTC members, 14 percent; nonmember participants, 9 percent.

A second allocation scheme was presented before a special meeting of the IATTC in 1973. It offered four criteria for allocating the allowable catch among IATTC member nations:

1. The ratio of each member's coastline length within the regulatory area to the distance from the center of that coastline to the center of distribution of yellowfin within the CYRA
2. The population of each member expressed as a proportion of the combined population of all members
3. The consumption of tuna taken from the CYRA by each member expressed as a proportion of total consumption from the CYRA
4. The CYRA catch of each member in the previous year expressed as a proportion of total CYRA catch in that year

An average of these proportions would have been computed for each country. This figure when applied to the overall yellowfin quota (less 10,000 tons reserved for nonmember nations to share) would have determined each member's national quota. Allocation by this method would have alloted 55 percent of the overall quota to the

United States, 25 percent to Mexico, 2 percent to Costa Rica, and 18 percent to the remaining member nations. Nonmember nations would not receive separate national allocations, even those with substantial adjacent resources and significant fishing industries (e.g., Ecuador).

Under any formula for allocating the total catch, it is quite certain that the two criteria of resource adjacency and historic participation in the fishery will account for the major share of the allowable catch. In fact, the entire quota could well be allocated on the basis of just these two criteria. If this were done in the eastern Pacific where RANs and non-RANs already participating in the fishery are fully capable of taking the entire overall quota, there would be no provision for any new non-RANs to enter the fishery. But if nations with legitimate interests in entering the fishery are excluded from the management program, they might elect to fish beyond 200 miles in disregard of the program. Because this could disrupt management to the extent that it might become totally ineffective, some consideration should be given to the possibility of setting aside a share of the allowable catch which could be taken by nations entering the fishery for the first time, as is done by ICNAF. However, allocation of tonnage for new entrants would pose a serious problem in the eastern Pacific. During 1977, fifteen nations (six RANs, nine non-RANs) fished in the eastern Pacific, with more showing an interest in entering. Suppose the yellowfin resource was to be allocated solely on the basis of resource adjacency and historical participation. To take an extreme case based on 1970–74 figures, suppose further that 57 percent (109,000 tons) was allocated to the ten RANs of the region based on the adjacent zone catch data given in Table 2 (i.e., the high allocations in Tables 3 and 5), and that these allocations were not subject to reduction. This would leave 43 percent (84,000 tons) to be divided among non-RAN historical participants, and one of these, the United States, already has enough vessel capacity to harvest the entire yellowfin catch available in the eastern Pacific. Obviously, if any part of this 43 percent share were to be set aside for new entrants, then this would reduce the allocations of historical participants and further exacerbate the problem of catch distribution. Although this is an extreme case, it is clear that no matter how it is handled, the new-entrant nation problem stands as a major obstacle to total resource allocation, especially since tuna can be harvested beyond 200 miles.

Besides determining the basis for total allocation, the nations involved in such a management approach would have to deal with several additional questions in organizing an effective system. One such question would concern the transferability of quotas. In discussing PAQ management, the alternative of treating allocations as nontransferable guarantees of access to shares of the catch was stressed over the alternative of treating them as transferable property rights. With guarantees of access, if a nation failed to harvest all or part of its allocation, then that amount would revert to the unallocated pool to be taken on a competitive basis. Such nations would receive some compensation in the form of increased net participant fee disbursements. With total allocation, however, nations would probably want to consider their allocations as property rights that they could transfer to other nations in exchange for cash or other considerations. Another question would be how to treat skipjack. Because skipjack are apparently not fully utilized in the eastern Pacific, there is no control on the level of harvest in the form of an overall quota. Country allocations for skipjack would be artificial under these circumstances. Therefore, total allocation schemes in the eastern Pacific, if they are to be considered, should initially apply only to yellowfin.

Problems relating to the variable geographic distribution of fishable concentrations of tuna and the resulting need to maintain accessibility to the resource over its entire range were discussed with regard to PAQ management in Chapter 8. Similar considerations would apply in the case of total allocation of the resource. Even though nations would be given allocations, there would be no assurance that they could harvest them unless they could fish wherever fish became available. Therefore, some system for licensing vessels to fish within juridical zones would be necessary for a functional management program. A system of participant fees would probably also be desirable in a management program based on total allocation. Management and enforcement costs could be paid for with participant fees just as in the case of PAQ management (see Chap. 8). Participant fees might also be used in recognizing RAN claims to special consideration based on their adjacency to the skipjack resource which would not initially be allocated.

In addition to the new-entrant problem already discussed, the total allocation approach has several additional drawbacks. Under total allocation, competition among nations is reduced because each nation can fish only until its quota is filled and then must stop. Though excessive competition for fisheries resources is generally considered somewhat wasteful, it does result in a high degree of technological efficiency. Indeed, competition provided the basic stimulus for the evolution of the modern tuna purse seiner, and competition is likely to be a major factor in bringing about changes to more efficient gear in the future. Hence, a possible shortcoming of total allocation is that it might tend to discourage technological innovation. On the other hand, even if tuna resources are fully allocated, some competitive stimulus for improvements in technological efficiency would still remain. For example, no one would suggest that there is a lack of competition or technological innovation in United States agriculture, even though land is completely allocated.

Total allocation could also have the effect of fixing a nation's catch in perpetuity. In the case of yellowfin in the eastern Pacific, a nation with an allocation could not increase its catch by merely increasing its effort for there would be no unallocated shares to be taken on a competitive basis. In certain respects this would be good: It would tend to discourage development of excess fleet capacity and would allow each nation to optimize the fishing strategy of its fleet. But in another respect this could be considered a drawback: Nations would be fixed in terms of future growth in the harvesting sector of their tuna industries. For example, a RAN such as Nicaragua would be allocated only a few thousand tons of yellowfin on the basis of adjacency and historic catches, hardly enough catch to maintain a single, large vessel. Yet Nicaragua has plans to develop her tuna fleet and by 1977 already had two vessels totaling more than 2,000 tons of capacity. These two vessels alone will require in excess of 5,000 tons of fish per year to operate profitably. With total allocation there would be no share of the yellowfin resource for Nicaragua to fish for on a competitive basis. This means that fleet expansion in Nicaragua would have to be curtailed, or else that her vessels would have to fish for other species or in areas outside the eastern Pacific. Another possibility would be for Nicaragua to negotiate for transfers of allocations from other nations.

A further drawback to total allocation in the eastern Pacific lies in the potential exclusion of nations from fisheries in areas where they are neither RANs nor historical participants. In a sense this is the new-entrant problem in reverse. Non-RANs take the major share of the catch in the eastern Pacific, and some of these nations already fish heavily (or may desire to do so in the future) in other oceans of the world where they are also non-RANs. If a world precedent is set for total allocation, then the major tuna fishing nations—Japan, the

United States, the Republic of Korea, Spain, the Republic of China, France—could conceivably be excluded from access to a major share of the potential world harvest. Exclusion from other oceans could also pose future problems for RANs such as Mexico which have developing fleets that are presently active only in their own region. The strategy of such nations, RANs and non-RANs alike, might well be to oppose total allocation.

When the advantages and disadvantages of totally allocating an overall quota are weighed against each other, it seems somewhat questionable that a successful agreement could be reached. Perhaps the most difficult obstacle would be to secure non-RAN agreement as to the size of their allocations. Even if such agreement could be secured, there would remain the serious problem of fishing by non-cooperating nations beyond 200 miles. Nevertheless, total allocation of the resources should certainly by given careful consideration.

11 Resource Allocation by Competitive Bidding

From time to time competitive bidding has been proposed as a possible method for determining who is to harvest a fishery resource. Though little has come of these proposals, it is not surprising that they are made because ample precedent exists for allocating natural resources by putting them up for bid. For example, in the United States companies bid competitively for the right to explore for oil and gas on tracts of land leased by state or federal governments. Auctioning of timber harvest rights in United States National Forests provides another example.

Management systems involving competitive bidding for harvesting rights have certain advantages, at least in theory. One advantage is that bidding eliminates the problem of determining an appropriate level for participant or licensing fees. Harvesters make this decision for themselves with their bids. Competition should drive bids up to close to the limit that more efficient harvesters could afford to pay, and this amount ought to exceed fixed fees that are based to any extent on the ability of less efficient harvesters to pay. Hence, revenues derived from the resource for distribution by an international management agency should tend to be maximized. Another significant advantage of bidding is that catch allocation and gear limitation problems would be neatly resolved by the harvesters themselves through their bids. Successful bidders would remain active in the fishery and share the catch among themselves, while vessels of unsuccessful bidders would be forced out.

With these advantages, a basic question is to what extent they are offset or perhaps outweighed by undesirable consequences of competitive bidding. Unfortunately, proponents of competitive bidding as a means for allocating a fishery resource have not generally gone into detail as to how an actual bidding system might be organized, and the practicality of bidding really cannot be evaluated without such knowledge. In what follows, an attempt will be made to identify and evaluate problems associated with various bidding systems and to suggest modifications that might effectively overcome some of the major objections. It will be assumed throughout this discussion that there is an international tuna management organization with the capability to determine appropriate harvest levels for tuna stocks under its jurisdiction. Harvesting rights to all or part of the allowable catch would be distributed through some sort of competitive bidding system (e.g., open bidding, sealed bidding, Dutch auction bidding, etc.), with successful bidders being guaranteed access to fish for their shares throughout the range of the stock to within 12 miles of coastal nations. Bidding would be administered by the international agency and proceeds would be used first to cover management costs, with the remainder than being distributed among coastal nations and harvesting nations just as outlined in Chapter 8. Distributions to coastal states on the basis of resource adjacency would provide partial recognition of their claims to special consideration.

In evaluating bidding systems, it is logical to begin with a very unrestrictive system. Assume that bidders compete for shares of a predetermined total allowable catch, and that any bidding entity can bid for any share of the catch. A bidding entity could be an individual, a corporation, a consortium, a nation, or a cooperating group of nations. A successful bidder could do as he pleased with his share of the allowable catch. He could harvest his share, transfer his harvesting right to any other entity, or even leave his share unharvested. The only requirement would be that successful bidders pay the international agency in a timely fashion, regardless of their plans for their shares. This system could foster purely preclusive bidding as countries attempted to drive one another out of the tuna fishery and companies tried to bankrupt one another. By bidding high enough, even to the point of sustaining short-term losses, the strongest-bidding entities could prevent weaker competitors from having access to the resource, the idea being to recoup any losses in subsequent years. It is hard to imagine such a system being acceptable to anyone except possibly those in the very strongest competitive positions.

To avoid some of the more extreme abuses of unrestrained competitive bidding, certain restrictions might be considered. A logical first step might be to ensure that the intent of all bidders was to fully harvest the shares of the catch that they bid for. This could be accomplished by requiring a successful bidder to make punitively large default payments for failure to fully harvest his share of the catch (or at least for failure to make in good faith an effort to do so). In the event an original bidder transferred his harvesting right, the new recipient would assume the harvesting responsibility and be liable for any default payments. This would eliminate bidding for catch shares with an intent to prevent their harvest. A successful bidder who had intended to harvest his share, but who for some reason discovered he could not or did not wish to do so, would have to transfer his harvest right to avoid default payments. To expedite such transfers, the agency could function as a broker in bringing together potential buyers and sellers. While default payments could eliminate some of the most extreme bidding practices, the competition for shares of the catch would remain intense, and weaker-bidding entities still might not survive.

If any entity can bid and if successful bidders can freely transfer their harvesting rights, it is likely that some speculators would bid purely in anticipation of later being able to transfer fishing rights to harvesters at a profit. While a default payment system would tend to inhibit such speculative bidding, especially by entities without harvesting capability, speculation could still be an important element. This speculative aspect should tend to maximize agency revenue because the number of prospective bidders would be maximized. But if it is considered desirable to reduce or eliminate speculative bidding, additional restrictions could be adopted. For example, bidding could be limited to only those entities possessing a harvesting capability. This would tend to reduce speculation, even if successful bidders were still allowed to transfer fishing rights. If rights were nontransferable, and a default penalty system was in effect, then only potential harvesters could bid, and the speculative element would be eliminated.

An alternative approach to the problems associated with unrestrained competitive bidding would be to allow only nations to bid for shares of the catch. Once determined by bidding, national shares would not be transferable between nations, and the internal allocation of such shares would be left up to the nations themselves. This approach might conceivably be acceptable to a developing nation with a small, perhaps nationalized, fishing industry, but in a nation with a large and highly competitive industry, the various interests and elements might well object to national bidding. The danger of preclusive bidding by nations with larger fleets would also be inherent in a national bidding system.

Instead of competitive bidding for shares of a predetermined total catch, suppose bidding is for licenses or the right to fish. The international agency could determine an appropriate level of gear for a particular fishery (e.g., a specified number of 1,000-ton purse seiners or their equivalent in vessels of other types and sizes); the vessels of successful high bidders would then fish competitively until a predetermined overall catch quota had been taken. As in the case of bidding for shares of the catch, licenses in a gear-limited fishery could be either transferable or nontransferable. The fact that successful bidders would fish competitively with no individual limit on their catches would probably be sufficient to insure that all bidders intended to harvest the resource, so default payments should not be necessary. However, restrictions on eligibility to bid might be desirable along lines considered in the case of bidding for catch shares. Possibly each successful bidder could be guaranteed some minimum catch as a form of insurance against boat breakdowns or other problems that might limit a vessel's fishing time prior to closure.

Bidding for licenses would have the advantage of dealing directly with the problem of effort limitation. Although both the total amount of gear and total catch would be controlled, such a system should actually be more competitive than one in which only total catch is controlled and bidding is for shares of that catch. This follows because the fishing itself would be competitive with no limitation on an individual vessel's total catch and no guaranteed catches (except perhaps for some minimal quantity in cases of hardship). In effect, bidding for licenses would result in a competitive form of limited entry.

Two basic systems for distributing resources purely by bidding have been considered: In one, total catch would be limited, and bidding would be for shares of that catch; while in the other, total catch and total amount of gear would both be limited, and bidding would be for licenses to fish with fishing itself on a competitive basis. In both cases stocks would be treated strictly as international resources with no special allocations for RANs. Because RANs with developing fisheries might find themselves at a considerable disadvantage in bidding against interests from nations with well-developed industries, it is likely that neither system would be acceptable to RANs desiring to increase their level of participation. It is natural then to consider how these bidding systems might be modified to provide recognition of RAN needs.

As in the case of PAQ management, assume that RANs are granted allocations related in some way to catches made within the 200-mile zones off their coasts. Each RAN working with the international agency could determine what part of its allocation it could harvest in a given year. This could range from none of it to all of it. For whatever portion they decided to harvest they could be required to pay a fixed RAN participant fee. The remainder of the total allowable catch, consisting of resources taken beyond 200 miles plus any unutilized RAN shares, could go into a common pool to be auctioned. The fixed RAN participant fee might be based on these bids and might logically fall somewhere between the minimum successful bid and the average successful bid. Bidding could be either for shares of a fixed total catch or for gear licenses. In the latter case, fees paid by RANs could be related to the number of standard units of gear required to harvest the catch to be taken under their allocations. Other details of bidding, including who should be eligible to bid, would have to be worked out, but RANs should be allowed to bid competitively for shares of the catch beyond their allocations. None of this should present insurmountable difficulties if the principle of bidding combined with national allocations is generally accepted. As in the case of PAQ management, proceeds deriving from bids and fixed RAN fees could be used first to pay for management costs; then a portion could be distributed to coastal states on the basis of resource adjacency; and finally the remainder could go to harvesting nations on the basis of their catches.

A system of bidding combined with RAN allocations based on resource adjacency would be similar in several important respects to the PAQ management system discussed in Chapter 8. Hence, it is not surprising that many of the problems that would have to be resolved in implementing a bidding-with-allocations system would be similar to problems discussed at length in considering PAQ management. In the eastern Pacific, these problems would include, among others, determining the method for closure of the fishery, developing a multispecies management policy for yellowfin (fully exploited) and skipjack (underexploited), phasing in RAN allocations, and controlling fleet size. Solutions to such problems might depend to some extent on actual bidding arrangements, but material presented in Chapter 8 and in Appendixes I–IV makes further detailed analysis unnecessary here. Suffice it to say that bidding could conceivably prove to be a viable means for distributing the catch, but probably only if combined with some form of adjacency-based allocations for RANs.

12   The Special Problems
of Porpoise and Billfish
Conservation

As pointed out in Chapter 4, an effective management body, in addition to developing an efficient system for collecting and analyzing the basic data necessary for the management of tuna resources, must also deal with other species taken during tuna fishing activities. Various species of bait fish constitute one such category, and in the 1950s and early 1960s the IATTC devoted substantial effort to bait fish studies. With the decline of bait fishing in the eastern Pacific, these studies have been phased out. However, bait fishing on a major scale continues in other ocean areas where conservation of bait fish remains an important concern.

A second example of a multispecies management problem involves the various species of porpoise captured in association with yellowfin tuna by purse seining in the eastern Pacific. Since 1959 this has resulted in considerable porpoise mortality, and, not surprisingly, there has been widespread concern over the effects of this mortality on porpoise stocks. This concern has led to national regulations that could have widespread consequences, both for the tuna fishery and for the evolving management program in the eastern Pacific. Because of its importance, it is worthwhile to explore the problem of fishery-induced porpoise mortality in considerable detail, taking into account biological, technological, and political aspects. The following sections will deal with the purse-seining technique for capturing tuna associated with porpoise schools and the magnitude of the resulting porpoise mortality; United States efforts, both legislative and technological, to resolve this problem; and progress toward resolving the porpoise mortality problem at the international level.

Still another multispecies management problem involves billfish, which are often exploited together with tuna in longline fisheries. Here the key problem is that there are two distinct groups of harvesters—commercial fishermen and sport fishermen—who hold sharply conflicting views concerning the proper objectives of management. The possibilities for resolving this conflict will be briefly considered in the final section of this chapter.

The Fishery on Associated Schools of Tuna and Porpoise

As has been noted, the eastern Pacific Ocean from the west coast of the Americas to about 150°W longitude supports a major tuna fishery. Prior to 1958 bait boats constituted roughly 80 percent of the vessel capacity operating in the eastern Pacific. With the development of economically more efficient purse seining made possible by the introduction of the hydraulic power block for hauling the net and nylon webbing, the composition of the fleet changed rapidly. In the mid-1970s, out of about 400,000 short tons of tuna taken annually in the eastern Pacific (mostly yellowfin and skipjack), roughly 300,000 tons were taken by purse seiners. Purse seiners in the eastern Pacific fleet range in size from small vessels that can carry about 100 tons of frozen tuna up to large and very modern super seiners capable of carrying as much as 2,000 tons of frozen tuna.

Another important factor contributing to the successful transition to purse seining was the fact that yellowfin tuna are frequently associated with certain species of porpoise, especially in warmer water areas off southern Mexico and Central America. In this respect the two most important species by far are the spotted porpoise *(Stenella attenuata)* and the spinner porpoise *(S. longirostris).* In the case of the former, the offshore spotted stock is most important, while the latter has two major stocks, the eastern spinner stock and the whitebelly spinner stock. The approximate known ranges of these stocks are indicated in Figures 8 and 9. It is worth noting that spotted and spinner porpoise are not species commonly seen in captivity. Also, although they occur along with yellowfin in other ocean areas, they are not commonly associated, and a porpoise fishery for tuna exists only in the eastern Pacific. Yellowfin also associate on occasion with at least six other species of marine mammals, including the common dolphin *(Delphinus delphis)* and the striped porpoise *(S. coeruleoalba).* Spotted, spinner, and striped porpoise and the common dolphin are all species whose snouts are elongated into a toothed beak. Technically speaking, all are considered to be dolphins. In keeping with widespread common usage, however, we use the generic term "porpoise" in referring to these species.

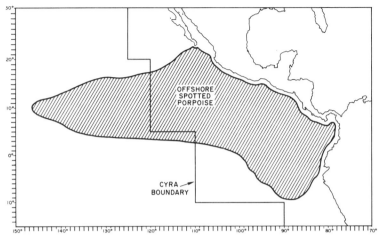

Figure 8. Approximate range of the offshore spotted porpoise stock. (Source: NMFS)

Fishermen in the eastern Pacific have capitalized on this tuna-porpoise association in two ways. First, because porpoise are larger and usually more active at the sea surface than tuna, fishermen often locate tuna by searching for porpoise and associated birds. Second, fishermen found that because of the strong behavioral bond between tuna and porpoise, if they can successfully encircle a school of porpoise within their net they are likely to capture tuna associated with the porpoise. This fishing method did not become important until the 1960s.

The ability of modern super seiners to harvest schools of yellowfin associated with porpoise has had a profound impact on tuna production in the eastern Pacific. It has enabled fishermen to take tuna further and further offshore from the traditional coastal fishing grounds. In fact, as vessels move further offshore the percentage of yellowfin taken in association with porpoise increases until in the area to the west of the CYRA it exceeds 95 percent. Also, on the average, larger and older yellowfin are available to the offshore fishery on porpoise. Because a unit of yellowfin recruitment can produce a greater sustainable yield if harvested at these larger sizes, offshore fishing makes good sense from a management and production point of view. For these reasons, and because of continued fleet growth, yellowfin production increased steadily through the 1960s and early 1970s.

Originally, after sighting a porpoise school associated with tuna (or birds flying over such a school), the fishing vessel would simply attempt to approach close enough to set its net around the school. Often, however, when the school was approached it sped up enough to make it impossible to catch. On other occasions the school would split up into several smaller groups and the vessel, if it released its net, would catch only part of the original school, or perhaps none of it.

Reacting to these problems, fishermen soon began to employ speedboats to herd the porpoise, thus increasing the likelihood of being able to set successfully on a school and also increasing the proportion of the school that could be encircled and retained. Herding with speedboats is effective because porpoise avoid them as well as their wakes. Typically, a seiner employs four or more speedboats, and the drivers are equipped with a citizen's band radio for receiving instructions from the captain. Generally, when a porpoise school is sighted (or birds over a school) the vessel approaches to determine the likelihood of tuna being present. If it appears high, the speedboats are launched, and the vessel resumes its pursuit of the school, with the speedboats following closely behind. When the vessel is sufficiently close to the porpoise school the speedboats accelerate ahead and approach the school. Following the captain's instructions, they encircle the school and attempt to herd it into a stationary and compact mass that the vessel can approach and encircle in its purse seine. When the net has been set, the speedboats are retrieved except for one or two that remain in the water to assist in porpoise release.

The porpoise that are caught have no market value, so they must either be released alive or, if they are killed, discarded at sea. Aside from humanitarian considerations, fishermen derive two benefits from releasing them alive. First, they will presumably associate with yellowfin in the future, thus enabling the capture of more tuna. Second, when porpoise are killed, valuable fishing time is lost in removing them from the nets. During the early days of the fishery, no satisfactory technique had been developed to release porpoise from the net after pursing. Fishermen commonly went into the water inside of the net after most of it had been brought aboard and removed as many porpoise as possible by pushing them over the corkline, but this was time consuming and dangerous because of sharks. Also, by that stage of the set many animals had already been killed. Consequently, the proportion of encircled porpoise being killed was relatively high.

In attempting to reduce porpoise mortality fishermen developed two effective approaches. The first to be adopted was the backing

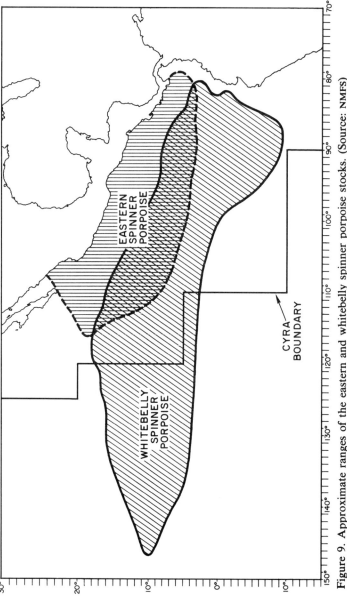

Figure 9. Approximate ranges of the eastern and whitebelly spinner porpoise stocks. (Source: NMFS)

down procedure, followed about a decade later by incorporation of the Medina safety panel into purse seine design. The backing down procedure was developed by United States fishermen in about 1960. Backing down begins after the seine has been pursed and from one-half to two-thirds of the net encircling the tuna and porpoise has been hauled aboard. Both ends of the net are secured. Then, at a critical time when the porpoise are near the far side of the net away from the vessel and the tuna are headed toward the vessel, the engine is shifted into reverse and power applied (usually about one-quarter to one-third throttle). As the vessel moves backward, the strain of water passing through the net causes the corkline at the far end of the net to sink a few feet below the surface, allowing the porpoise to escape (see Fig. 5). In effect, the net is pulled out from under the porpoise milling at the surface. As they are forced out of the net, most porpoise realize immediately that they are free and react by racing away, leaping repeatedly from the water as they go. (Porpoise avoid the corkline and will not normally jump over it, although they are physically quite capable of doing so.)

The backing down maneuver, unless carefully timed, can result in tuna as well as porpoise escaping from the seine. To prevent this, one or more speedboats are stationed in the release area with their engines off. The men in these speedboats signal when the tuna are clear of the release area and headed back toward the boat, at which time power is applied to sink the corkline in the release area. When the tuna turn back toward the release area, the throttle is closed, and the men in the speedboats pull the corkline back to the surface to prevent escape of tuna. In this manner, backdown can proceed in stages until as many porpoise as possible are freed from the net. The men in the speedboats also assist porpoise out of the net in the backdown area, and after backdown they can work their way down the corkline to assist any live porpoise remaining in the net.

Despite these efforts to save porpoise, many animals still became entangled and died, especially in areas where they could not be reached and when they were concentrated in the apex of the backdown area. To alleviate mortality due to entanglement during backdown, fishermen in 1971 began to replace the uppermost strip of 4¼-inch-mesh webbing in the release area with a 120-fathom strip of 2-inch-mesh webbing. (A strip of net is approximately 5½ fathoms in depth.) This finer mesh insert is often referred to as the Medina strip or safety panel after Captain Harold Medina, who first developed the concept. Most of the contact that porpoise have with the net occurs during backdown in the safety panel area, and fewer porpoise entangle their snouts and flippers in its smaller meshes, thus reducing mortality. Because of its obvious benefits, the Medina or safety panel modification was rapidly incorporated into the gear of all United States vessels and many vessels fishing under other flags.

The incorporation of safety panels into the seines of United States vessels was further stimulated by enactment of the Marine Mammals Protection Act (MMPA) on October 21, 1972. Although the MMPA formally took effect on December 21, 1972, under its terms the incidental taking of marine mammals in the course of commercial fishing operations was allowed to continue until October 21, 1974, provided that the fishing techniques and equipment which were used would produce the least practicable hazard to marine mammals. Clearly the adoption of safety panels by many United States vessels represented a significant step in reducing the hazard to porpoise caught with tuna, as called for by the MMPA.

Before we further explore the requirements of the MMPA and events ensuing from its passage, we should examine the impact of the expansion and development of the purse-seine fishery in terms of porpoise mortality. Annual mortality estimates for the international fleet from 1959 onward are given in Table 8 for the three stocks of porpoise most heavily involved in the fishery: offshore spotted porpoise and eastern and whitebelly spinner porpoise. These estimates are based on NMFS marine mammal observer data and IATTC tuna logbook data. (The IATTC logs, while not containing information on the number of porpoise killed, do enable determination of the number of sets made on tuna associated with porpoise during past years.)

Because of the major assumptions made in developing these mortality estimates, they can only be considered as rough approximations. In particular, they do not take into account any mortality that occurs outside the net such as mortality owing to fatigue or to injuries sustained in the net. These estimates are, however, the best information available on historical porpoise kills and can be interpreted in light of the development of the fishery for tuna associated with porpoise.

Significant porpoise mortality began in 1959 when the fleet first began to adopt porpoise fishing techniques on a meaningful scale. Mortality is estimated to have increased rapidly as more and more vessels shifted to porpoise fishing (1960–61). Soon, however, the an-

TABLE 8
ESTIMATED PORPOISE KILLS FOR PRINCIPAL STOCKS BY VESSELS OF
ALL FLAGS IN THE EASTERN PACIFIC OCEAN, 1959–77
(Thousands of animals)

	Species or stock			
Year	Offshore spotted	Eastern spinner	Whitebelly spinner	Total†
1959	68	40	0	109
1960	534	278	0	853
1961	446	232	0	713
1962	106	55	0	169
1963	133	69	0	213
1964	255	133	0	407
1965	297	155	0	475
1966	281	146	0	449
1967	195	101	0	311
1968	164	85	0	262
1969	331	158	14	529
1970	308	134	26	492
1971	185	99	31	315
1972	273	44	21	338
1973	120	44	30	194
1974	75	22	18	115
1975	106	26	39	171
1976	85	9	40	134
1977*	16	2	5	23

SOURCE: NMFS
*For 1977 the estimated U.S. kill through December 11 is given.
†Totals include kills of other species.

nual estimated mortality dropped sharply, probably reflecting the efforts of fisherman to save porpoise, especially utilization of the backing down technique (1962–63). Estimated mortality then increased again as the number of vessels fishing on porpoise continued to increase and as the fishery moved further offshore into areas where fishing on porpoise predominated. During the 1964–68 period, estimated mortality fluctuated around a mean annual level of about 380,000 animals with some of the variation between years probably reflecting changes in the availability of smaller yellowfin ("schoolfish") and skipjack not associated with porpoise. For example, 1967 was an exceptionally good year for skipjack in the eastern Pacific. Another factor affecting the porpoise kill was the regulation of the yellowfin catch through adoption of overall yellowfin quotas beginning in 1966. In 1969 the fishery expanded into the area west of the CYRA for the first time. In this area yellowfin are almost always harvested in association with porpoise, which may account in part for the increased estimated porpoise kills for 1969 and 1970. Also, it is believed that in 1969 the fishery for the first time began to impact upon the whitebelly spinner stock which is distributed offshore. From 1971 through 1977 a generally declining trend in the porpoise kill estimates can be noted, reflecting further improvements in fishing gear and techniques, especially the incorporation of safety panels into purse seines. In addition, adaptation and learning by the porpoise themselves have probably contributed to mortality reduction. It is reported that porpoise have become increasingly difficult to capture, and that, when they are captured, their behavior in the net is less panicked than in earlier years. Finally, the decline in mortality in part reflects the impact of the MMPA on the United States fleet, especially in 1977 when a major reduction in kill was achieved that is discussed in more detail later in this chapter.

The MMPA and Ensuing Regulations

With enactment of the MMPA in 1972, the United States government assumed a major role in the drive to reduce porpoise mortality. In passing this act, Congress stated in its findings that "certain species and population stocks of marine mammals are, or may be, in danger of extinction or depletion as a result of man's activities"; and that, as a matter of policy, "they should not be permitted to diminish below their optimum sustainable population[s]." It also directed that "measures should be immediately taken to replenish any species or population stock which has already diminished below [its optimum sustainable] population."

Obviously several of the terms incorporated in this formulation of policy required further definition, and this was to some extent provided in the act. For example, "population stock" was defined as "a group of marine mammals of the same species or smaller taxa in a common spatial arrangement, that interbreed when mature." "Depletion" was defined as the condition which occurs when "the number of individuals within a species or population stock . . . has declined to a significant degree over a period of years . . . [or] is below the optimum carrying capacity for the species or stock within its environment." "Optimum carrying capacity" was then defined as "the ability of a given habitat to support the optimum sustainable population." And finally, "optimum sustainable population" (osp) was defined "with respect to any population stock, [as] the number of animals which will result in the maximum productivity of the population or the species."

Following through this series of definitions, note that a stock must be termed "depleted" if it falls below osp. However, a stock that is depleted in this strictly legal context is not necessarily in danger of becoming extinct or even declining. Furthermore, even though the term "maximum productivity" is not defined in the mmpa, the definition of osp would not appear to preclude some killing of porpoise as long as viable and productive stocks are maintained. However, another section of the mmpa contains the following language: "In any event it shall be the immediate goal that the incidental kill or incidental serious injury of marine mammals permitted in the course of commercial fishing operations be reduced to insignificant levels approaching a zero mortality and serious injury rate." Even though the term "approaching zero" is left undefined, this goal appears to be somewhat in conflict with the previously stated objective.

As noted earlier, the mmpa did not take effect with respect to commercial fishing operations until October 21, 1974. After that date marine mammals could be taken (i.e., harassed, hunted, captured, or killed) only under regulations prescribed in permits issued by the secretary of commerce, who is obligated to adhere to detailed procedures spelled out in the mmpa in issuing permits. Under these provisions, general permits for the taking of porpoise in the course of purse-seining operations have been awarded to the American Tunaboat Association on behalf of the fishermen. As a condition of being allowed to fish under these general permits, vessels have been required to utilize gear meeting certain specifications and to maintain detailed records of their fishing activities. The first general permit covered the period from October 21, 1974, to the end of 1975. The second general permit covered 1976. In addition to gear and record keeping requirements, selected vessels fishing under this permit were required to carry government observers. The nmfs also reserved the right to establish a kill quota for any porpoise stock and to prohibit further setting on stocks for which the quotas had been reached.

Environmental organizations vigorously protested the issuance of these general permits, and in 1974 they brought suit against the secretary of commerce and other federal officials, charging that the permits had been issued improperly for a number of reasons. After many months of litigation, Judge Charles R. Richey of the U.S. District Court, District of Columbia, on May 11, 1976, rendered his critically important decision. He found that the NMFS had failed to fully discharge its obligations under the MMPA and ordered that the taking of porpoise should cease effective May 31, 1976. This ruling, which threatened to seriously cripple the United States tuna fishing industry, triggered a series of legal maneuvers which delayed the implementation of Judge Richey's order.

Perhaps the most significant result of the Richey decision was the establishment of porpoise kill quotas. The first such quota was for a 1976 kill of 78,000 animals without reference to specific stocks. This quota was promulgated on June 11, 1976, and on October 22 it was estimated that the quota had been filled. After some delay resulting from legal actions, porpoise fishing was prohibited beginning on November 11 for the rest of 1976.

When the 1976 porpoise quota was set, it was clear, based on Judge Richey's ruling, that future quotas would have to be established on a stock-by-stock basis in order to comply with the requirements of the MMPA. Therefore, the NMFS organized a workshop on the assessment of porpoise stocks involved in the eastern Pacific yellowfin tuna fishery. A group of scientists, including experts in population analysis, was assembled for this workshop, which was held in July 1976 at the NMFS Southwest Fisheries Center in La Jolla, California. The charge to the workshop participants, as stated by the director of the NMFS, was "to produce estimates of existing population levels, OSP (optimum sustainable population) levels, and the impact of incidental taking on those levels for each and every species or population stock of small cetaceans involved in the U.S. yellowfin tuna purse seine fishery."

Before the workshop could address the above objectives, some clarification of the previously mentioned semantic difficulties embodied in the MMPA was necessary. Because OSP is defined in terms of maximum productivity, the first step was to agree on the type of productivity intended in the MMPA. It was agreed that for the purposes of the tuna-porpoise problem "productivity" should be interpreted as net productivity (the difference between the numbers of births and deaths from all causes) rather than gross productivity (the number of births) or productivity per unit population (the birth rate per individual).

With the net productivity concept in mind, the conventions of usage adopted by the workshop can best be explained by making reference to Figure 10. In theory, a population can be sustained at any average level between the maximum sustainable population (MSP) that is characteristic of an unexploited population and some minimum viable population (Fig. 10a). At MSP the number of births must equal the number of deaths on the average, and net productivity must be zero. This is also obviously true of a minimum viable population or if a population is driven to extinction. Therefore, maximum net productivity (MNP) must lie at some intermediate level. While MNP estimates are not available for porpoise, population models used in studies of other marine mammals (e.g., fur seals and both baleen and sperm whales) led workshop participants to conclude that any porpoise stock whose abundance is less than 50 percent of MSP is probably below the MNP level, while any stock at much more than 70 percent of MSP is probably above the MNP level (Fig. 10b).

Recalling the MMPA definition of OSP cited earlier which related OSP to "maximum productivity," it might seem logical to equate OSP with MNP. However, the workshop participants did not choose this course. Instead they introduced the concept of a range within which OSP must fall. The upper limit of the OSP range is MSP, and the lower limit is MNP, which probably falls somewhere between 50 and 70 percent of MSP (Fig. 10c, 10d). Combining this OSP range concept with the concept that a stock, by definition, is depleted if it falls below OSP, one can conclude that a stock is depleted if it falls below 50 percent of MSP (assuming, of course, that MNP is not below this level). At any level between 50 and 70 percent of MSP no clear-cut conclusion is possible owing to uncertainty concerning the MNP value. Above 70 percent of MSP a stock would be considered as falling within the OSP range.

A major reason for discussing the NMFS workshop interpretation of the MMPA is that its conclusions concerning the meanings of productivity, OSP, and stock depletion guided the United States government in its subsequent actions pertaining to porpoise, particularly in setting kill quotas. Despite the efforts of workshop participants, however, some ambiguity remains in the terminology of the MMPA and interpretation of it. If the MMPA requirement that the incidental kill must be reduced to an insignificant level approaching a zero mortality rate is interpreted to mean that fishery related mortality must be eliminated entirely, then this would imply that OSP must equal MSP and that a

a. RANGE OF SUSTAINABLE POPULATION
    SIZES EXPRESSED AS PERCENT OF
    MAXIMUM SUSTAINABLE POPULATION (MSP).

b. FOR ITS TERMS OF REFERENCE, THE NMFS
    PORPOISE WORKSHOP AGREED TO ASSUME
    THAT MAXIMUM NET PRODUCTIVITY (MNP)
    FELL BETWEEN 50% AND 70% OF MSP.

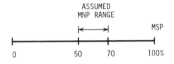

c. IT IS HYPOTHETICALLY POSSIBLE TO
    DETERMINE THE SPECIFIC VALUE OF
    MNP, ALTHOUGH THIS HAS NOT YET
    BEEN DONE FOR ANY PORPOISE STOCK.

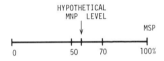

d. ACCORDING TO THE NMFS INTERPRETATION,
    THE OPTIMUM SUSTAINABLE POPULATION (OSP)
    FALLS WITHIN THE RANGE DEFINED BY MNP
    AND MSP.

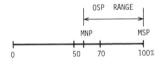

Figure 10. NMFS interpretation of the optimum sustainable population concept as set forth in the MMPA

virgin stock must be maintained. But with existing technology, it is not possible to eliminate porpoise mortality totally without also eliminating seining for porpoise-associated tuna and greatly reducing the yellowfin catch. Hence, some OSP value below the MSP level seems more realistic. But if a stock is to be maintained at a lower OSP level, this could be construed as implying that some porpoise should be killed if the stock rises above OSP, a conclusion that is obviously counter to the intent of the MMPA. This type of confusion could be eliminated if under the existing law OSP were to be set equal to either MNP or the existing stock size, whichever is greater, with the guiding policy being to increase porpoise populations to the extent that is technologically and economically feasible.

Having formulated its working definition of the OSP concept, the

NMFS workshop proceded with its analysis of the status of individual porpoise stocks in the eastern Pacific. Basically this involved estimating the kill levels, estimating the population sizes, and then combining this information with estimates of the gross reproduction and natural mortality rates to evaluate the impact of porpoise fishing within the context of the OSP concept.

Fishing mortality estimates have already been discussed for the three main stocks involved in the eastern Pacific fishery (see Table 8). To make population estimates, the NMFS utilized a line transect approach which involved sampling the density and size of porpoise schools within strata of known area. This was accomplished by conducting aerial and/or shipboard searchs over a presumably representative portion of each stratum. The size of each school encountered was estimated and the data for all strata were combined to arrive at an estimate of total stock size. Transect data were available from two separate United States sources: an aerial survey conducted in winter 1974 and shipboard observer data from the winter and spring periods of 1974, 1975, and 1976.

In either an aerial or a shipboard survey, estimates of school size could be biased or determinations of species composition incorrect. Bad weather could adversely affect observations and the observers themselves could differ in their ability to detect schools. Also, the size of a school might affect its probability of being detected. Population estimates generated from surveys apply only to the area surveyed, and if significant portions of a stock's range are not covered, this fact must be taken into consideration. For example, the 1974 aerial survey covered only the shoreward portion of the range of some stocks owing to aircraft limitations. Lack of complete knowledge concerning the ranges of certain stocks presents additional difficulties.

Most of the problems mentioned so far could apply to either aerial or shipboard surveys. In addition, there are problems that are unique to the latter. If commercial fishing vessels are employed, the line transect samples are not likely to be random. Instead, they will be concentrated in areas of porpoise (and tuna) abundance. Also, there is some evidence that porpoise react to avoid fishing vessels while they are still a considerable distance away. Such avoidance behavior could very well reduce the probability of a school being sighted. To estimate the amount of area being searched, the effective width of the searching path must be known. This requires determination at the time of sighting of the distance and bearing to the porpoise school with respect to the line transect course. Possibilities exist for introducing bias in measuring either or both of these quantities from shipboard, but in an aerial survey there should be much less bias.

In spite of these line transect sampling problems, which could result in either overestimates or underestimates of the porpoise stock sizes, workshop participants were still faced with arriving at some single figure for each important stock. Considering all factors, and examining the available data in several ways, they concluded that the 1974 aerial survey could probably be considered more reliable than the 1974–76 shipboard surveys alone. Because of the limited aerial coverage, however, they of necessity had to combine aerial and shipboard observer data in arriving at their final population estimates. As of the beginning of 1974, these estimates in numbers of animals were as follows: offshore spotted porpoise, 3.5 million; eastern spinner porpoise, 1.2 million; and whitebelly spinner porpoise, 0.4 million.

These population estimates are highly dependent on the treatment of the various possible sources of error noted earlier. The workshop participants, because they were working within the context of the MMPA, tended to adopt the philosophy of erring on the conservative side if they erred at all. It is logically quite possible to arrive at different estimates by adopting a different, but still reasonable, set of assumptions. For example, the IATTC, in a 1977 review of the tuna-porpoise problem, arrived at generally higher estimates of porpoise populations, even though they utilized essentially the same aerial and shipboard survey data. The differences in this case were mostly attributable to the treatment of the observations used to estimate searching path width.

With population estimates and historical kill estimates in hand, workshop participants proceeded with their assessment of the effect of fishing on porpoise stocks. The approach adopted was reasonably straightforward. A simple model was developed for calculating population size estimates for years both prior to and following January 1, 1974. Other inputs were the mortality estimates mentioned earlier, reproduction rates, and natural mortality rates. Little or no information pertaining to the last two parameters was available for eastern Pacific stocks, so participants were forced to rely on studies made in other ocean areas as well as on their own judgment.

Population sizes were back-calculated to the inception of fishing for each stock and also projected to 1976. (In order to examine a range of possible values for input parameters, several cases were considered.) These calculations enabled comparison of the 1976 stock levels with the unexploited stock levels as shown in Table 9. The conclusions given in this table are based on the agreement among workshop participants that the lower bound of the OSP range coincided with with MNP and that this value probably fell somewhere between 50 and 70 percent of MSP. The probable effects of future incidental mortality levels on the various stocks were then considered. Because of the considerable uncertainty as to assumptions made in calculating 1976 population levels, the participants determined several kill levels for each stock, including a kill level at which they were "virtually certain" that the stock would increase in size.

All of the preceding findings were incorporated into the final workshop report, which served as a basis from which the NMFS prepared a set of recommended United States kill quotas for 1977. The NMFS set as its objective to stay within the kill levels which the workshop had determined would be virtually certain to result in population growth. To arrive at proposed United States kill quotas, they subtracted the expected non–United States kill from the virtually certain kill levels. On this basis, the proposed quota for offshore spotted porpoises turned out to be 21,800 animals. No whitebelly spinner quota was proposed owing to the level of the expected non–United States kill. For the eastern spinner stock, the virtually certain kill level exceeded the expected non–United States catch. Nevertheless a zero-kill limitation was proposed because this stock was considered depleted in the sense of being possibly below the OSP range. Quotas

TABLE 9

CONDITION OF PORPOISE STOCKS IN 1976 AS COMPARED TO THEIR
UNEXPLOITED SIZES

Porpoise stock	1976 population expressed as percentage of unexploited population size	Conclusion
Offshore spotted	48 to 82%	Most probably within lower end of OSP range
Eastern spinner	37 to 75%	Either below or close to lower end of OSP range
Whitebelly spinner	63 to 78%	Within OSP range

SOURCE: NMFS

totaling 8,118 animals of other species were also proposed.

These proposals, together with other regulations pertaining to gear and fishing procedures, were then the subject of a public hearing presided over by an administrative law judge. In addition to focusing on matters concerning the porpoise stocks and the impact of incidental mortality on these stocks, the hearing also addressed the economic and technological feasibility of implementing the proposed quotas and other regulations. After weighing the voluminous testimony presented at this hearing by individuals representing organizations holding a variety of positions with regard to the tuna-porpoise problem and the proposed regulations, the NMFS promulgated its final regulations for 1977. These included kill quotas of 43,090 offshore spotted porpoise and 7,840 whitebelly spinner porpoise, quotas which were significantly higher than those originally proposed. The whitebelly spinner quota was further increased to 11,219 animals later in 1977 in light of newly available scientific evidence concerning the size of this stock. No quota was established for the eastern spinner porpoise, in keeping with its being treated as a possibly depleted stock. It was decided, however, that an accidental kill (in contrast to an incidental kill) of eastern spinner porpoise could be allowed in cases where eastern spinners were found to be intermingled with other types of porpoise after initiation of a set. But in no case was it permissible to set on a porpoise school known in advance to include eastern spinners. For other porpoise species, quotas totaling 8,120 animals were also established.

Fishing under these quotas, the United States fleet achieved a substantially lower porpoise kill rate in 1977 than in any other year since the development of the technique of purse seining for tuna associated with porpoise. Based on reports through December 11 by NMFS observers who accompanied 42 percent of all trips by United States vessels authorized to fish on porpoise, the estimated United States kill was 15,979 offshore spotted porpoise, 4,598 whitebelly spinner porpoise, and 5,900 porpoise of other types, well within the established quotas for all major porpoise stocks. The overall kill rate was 0.25 animals per ton of yellowfin caught by fishing on porpoise, down sharply from the 1976 figure of 1.14 animals per ton of yellowfin. This reduction can be attributed to conscientious efforts by fisherman to save porpoise and to improvements in fishing technology which will be described in the following section.

While the greatly reduced 1977 kill rate justifies guarded optimism concerning reduction of porpoise mortality, it is important to realize that fishing conditions encountered during 1977 may not be representative of other years. For example, the distribution of fishing effort was atypical in 1977 due to a two-month spring tie-up of most of the United States purse-seine fleet. Because of this tie-up, closure of the fishery in the CYRA was significantly delayed, and United States vessels continued to operate there rather than to the west of the CYRA where the porpoise are more naïve with respect to purse seining and perhaps more susceptible to mortality.

Looking to the future, steadily declining United States quotas have been established for the years 1978, 1979, and 1980 (Table 10). In establishing these quotas primary consideration was given to the technological and economic feasibility of achieving them without an excessive adverse impact on the fishing industry. In this regard, it can be noted in Tables 8 and 10 that the 1977 kills of offshore spotted and whitebelly spinner porpoise were within the 1978–80 quotas. But because the distribution of fishing effort in 1977 was atypical, there is no real assurance that future quotas can be achieved without restrictions on porpoise fishing. Even assuming that the successful reduction in 1977 can be continued in 1978–80, there could be some problems staying within the quotas for some of the less important stocks, most notably the southern stock of the common dolphin. For this stock the 1977 United States kill through December 11 of 2,873 animals (NMFS estimate) falls within the 1977 quota of 5,600 animals, but it substantially exceeds the 1978–80 quotas of 1,700, 1,400, and 1,000 animals, respectively. Although about 10 percent of all porpoise sets are made on pure schools of common dolphin (belonging primarily to the south-

TABLE 10

ANNUAL UNITED STATES PORPOISE KILL QUOTAS BY STOCK FOR 1978–80

Porpoise stock	Annual United States kill quotas		
	1978	1979	1980
Offshore spotted	35,500	28,400	21,300
Eastern spinner	0	0	0
Whitebelly spinner	13,500	10,800	8,100
Others	2,945	2,410	1,750
Total	51,945	41,610	31,150

ern stock), only about 4,500 tons of yellowfin are taken in these sets. This is less than 4 percent of all yellowfin taken on porpoise in the eastern Pacific. It seems probable that any losses resulting from an early cessation of setting on common dolphin schools could be made up by fishing on other types of porpoise schools.

As long as tuna are caught in association with porpoise by purse-seining methods, some mortality will occur. Hence, unless a major breakthrough is made in fishing technology, it is clear that further quota reductions must, at some point, begin to limit tuna production. But if the tuna industry can succeed in meeting the 1978–80 quotas, there will be no threat to the maintainance of large and viable porpoise stocks. Thus, further quota reductions, if they are to unattainably low levels, would seem difficult to justify, except on humanitarian grounds.

Development of Fishing Technology to Reduce Porpoise Mortality

Fishing-induced porpoise mortality could be eliminated by prohibiting the use of purse-seine gear in fishing for yellowfin tuna associated with porpoise. There is little likelihood, however, that all of the nations fishing in the eastern Pacific would agree to this prohibition because purse seining is presently the only economically efficient means of harvesting tuna in the offshore portions of the eastern Pacific. Bait fishing has never been successful in these areas, and while there is some longline fishing by the Republic of Korea, Japan, and the Republic of China, production is low and operations are economically marginal. Therefore, it is unlikely that purse seining in these areas can be replaced by bait fishing or longlining, and prohibition of purse seining on porpoise associated with tuna could result only in greatly reduced yellowfin catches. Although porpoise would be protected, the production of tuna would be far below what the stocks could support on a sustained basis.

If existing fishing methods that do not harm porpoise cannot replace purse seining, then to make further progress research must be directed toward development of new types of porpoise-saving gear and fishing techniques. Ideally, such research should lead to a fishing technique that would be economically more efficient than the present purse-seining technique. If this were the case, captains and boat owners would logically want to adopt the new methods quickly. This would also be true if the new technique was economically equivalent to or even somewhat less efficient than existing methods, provided the choice was between changing over or not fishing at all. However, if it would be unprofitable to fish by some proposed new method, then no one could afford to change. The key point is that the economics of the fishery must always be borne in mind as well as the welfare of the porpoise. Any changes in fishing technology should both benefit the porpoise and reward the fishermen. Fortunately, much recent gear research seems promising in both respects.

An ideal solution to the porpoise mortality problem would be to develop a totally new method for efficiently harvesting tuna that would otherwise be taken in association with porpoise by purse seining. Unfortunately, it seems unlikely that any radically new method can be developed in the near future that would be efficient enough to compete economically with purse seining. It is highly improbable, however, that porpoise mortality can ever be totally eliminated in a purse-seining operation, so such a breakthrough may be deemed essential by those who advocate no killing of porpoise whatsoever. Such a breakthrough would almost certainly require a long-term gear research and development program that would have to be substantive, broad ranging, imaginative, and well funded to have any chance of success. Revolutionary approaches would have to be studied and evaluated. These might include use of large floating traps, electrofishing devices, and/or fish aggregating devices.

Perhaps the most promising of these concepts is development of an effective fish aggregating device. Smaller yellowfin tuna ("schoolfish") are often associated with floating objects ("logs"), especially in the nearshore area. Fishermen catch these schools by setting their nets around the floating objects. In offshore areas yellowfin are on the average larger and are mostly captured in association with porpoise. If these larger tuna could be attracted to some type of floating device, instead of associating with porpoise schools, they could be harvested just like schoolfish. A floating device might broadcast sounds mimicking those made by porpoise schools, or it might dispense substances whose odors were similar to those associated with tuna food items or with porpoise schools. Lights might also be used to attract species that tuna feed on. Because porpoise schools are actively moving entities, a mobile floating device might aggregate tuna more effectively than a passively drifting one. Hence the device might be self-propelled, perhaps being equipped with sails. It could also be equipped to detect the presence of tuna aggregating beneath it and to signal the deploying purse seiner.

One attractive aspect of the aggregating device concept is that it is based on observed tuna behavior. Another is that it would utilize existing purse-seine gear, and thus its adoption would be economically less disruptive than switching to an entirely new fishing technique. Although the aggregating device concept is interesting, it is important to emphasize that chances for a real breakthrough of this nature are speculative. In the past fishermen have taken logs with them to the fishing grounds and deployed them in the hope of aggregating tuna, but they have had scant success. Some work on developing aggregating devices has also been undertaken by the Japanese in the western Pacific, but results have not been encouraging. On the other hand, there has been some success in the Philippine Islands in attracting tuna to anchored rafts where they are harvested by pole and line. Clearly there is something unique about a natural "log ecosystem" or a porpoise school that attracts tuna, and this unique "something" may be very difficult to define and effectively reproduce.

Quite a different approach to avoiding the capture of porpoise in purse seines would be to somehow separate the tuna from the porpoise when associated schools are encountered so that the tuna could be encircled without capturing the porpoise. Although little research has been undertaken in this area, some possible separation techniques have been suggested that involve attracting or herding either the tuna or the porpoise. For example, it is conceivable that tuna could be separated from porpoise by chumming with live bait or use of food odors. Another possibility is that sound might be used to attract or to herd the fish. However, in view of the strong bond between tuna and porpoise and the fact that most porpoise flee from approaching fishing vessels, the concept of separating associated schools of tuna and porpoise immediately prior to setting the purse seine seems unpromising.

From what has been said so far, it would appear, at least in the near future, that the objective of developing more effective methods for releasing captured porpoise from the purse seine seems more attainable than the objective of not capturing porpoise at all. In fact, it may well be possible to achieve further significant reductions in mortality levels while maintaining the present tuna harvest level. It is even conceivable that purse seining could become more efficient as a result of improvements growing out of a desire to minimize porpoise mortality.

The United States is the nation that has placed the greatest emphasis on developing technology aimed at enabling porpoise to be released from purse seines unharmed. Nor surprisingly, the United States fishermen themselves, by virtue of their vast experience in fishing with purse-seine gear, have played a major role in developing release methods. Indeed, they have been responsible for the two major advances described earlier, the backing down technique and the Medina safety panel. It seems certain that their knowledge, skill, practical experience, and intuition will continue to play an important role in improving fishing techniques.

The NMFS has also played an important role and has experimented with a number of fishing techniques and gear modifications designed to reduce porpoise mortality. One general problem has been the high expense of modifying and testing gear. Another hindrance has been lack of vessel time to test various ideas. To alleviate this problem somewhat, the IATTC has provided special 1,000-ton yellowfin allocations to the United States in 1975, 1976, and 1977 to enable vessel charters for porpoise research. In addition, vessel owners and captains have cooperated by making their vessels available for testing new fishing techniques and gear modifications. Nevertheless, time has not always been sufficient to fully test the efficiency of modified gear or changes in fishing methodology. The United States government also supports research related to the tuna-porpoise problem through the National Science Foundation (Research Applied to National Needs) and the Marine Mammal Commission, which make available contractual funding.

Finally, the United States tuna industry has played an active role in tuna-porpoise research. In 1975, with participation of producer, processor, and labor segments, the tuna industry established the Porpoise Rescue Foundation. The objective of this organization is to foster, promote, and fund the development of porpoise-saving techniques, not only through modification of gear and fishing strategies, but also through study of the porpoise and tuna themselves and through extension activities such as sponsoring workshops on porpoise-saving techniques. In the past, however, the available funding has been limited and not sufficient for chartering all of the needed vessel time. In 1977 the tuna industry established the United States Tuna Foundation to work on a wide variety of problems, including the porpoise mortality problem. The Porpoise Rescue Foundation now serves as the research arm of this body and receives funding through it. The Tuna Foundation has also provided funds sufficient to charter a modern purse seiner for the entire year of 1978 to carry

TABLE 11
Summary of Porpoise-Saving Concepts Developed by NMFS and United States Tuna Fishing Industry through 1977

Concept	Description	Result
I. Proven or promising concepts		
1. Backing down	Partially hauled net is towed backward through the water causing far end of corkline to sink, which releases porpoise	A basic and effective technique; now a standard procedure in purse seining on porpoise
2. Medina panel (safety panel)	A smaller-mesh panel (originally 2-inch mesh) at the apex of the backdown area which reduces porpoise entanglement	A basic and effective concept; now a standard part of purse-seine design, though in modified form (see Nos. 3 and 4 below)
3. Porpoise aprons and chutes	Addition of webbing on top of safety panel to form a shelf or ramplike release passage	Apron design has evolved through several versions. Recent versions seem very promising, and are being adopted in the fishery. Apparently facilitates porpoise release
4. Use of fine mesh webbing	Use of 1¼-inch mesh in aprons and considerably enlarged safety panels to minimize porpoise entanglement	Important improvement over original larger-mesh Medina panel; rapidly being adopted by the fleet

5. Use of speedboats	Tow on net prior to backdown to prevent net collapse; drive porpoise away from danger areas; hand rescue porpoise during and after backdown	Basic and effective techniques; now standard procedures for most vessels purse seining on porpoise
6. Porpoise rescue raft	Manned one-man inflatable raft in backdown area; observes porpoise and tuna and signals captain to begin or continue backdown; assists in hand rescue	A promising technique for increasing number of zero-kill sets

II. Less promising concepts

7. Sonic repulsion	Pounding on bamboo poles, 30-kHz sound, and feeding killer whale sounds were tested	Desired directional response not achieved
8. Pneumatic porpoise gate	Inflatable and deflatable rubber tube replacing a section of corkline allows net to be lowered and raised	Prototype tests unsatisfactory, operational complexities; concept abandoned
9. Skimmer net	Large lampara-type net used to try to skim or guide porpoise out of seine	Porpoise avoided skimmer net; concept abandoned
10. Dual backdown channels	Tried to seal tuna in stern side loop while porpoise were released from other channel	Unable to separate tuna and porpoise adequately; concept abandoned
11. Fluorescent dye packets	Change in dye diffusion pattern indicates subsurface currents that could cause problem sets	Dye too diffuse for interpretation; concept abandoned

TABLE 11 (continued)

Concept	Description	Result
12. Current ribbons	Weighted fluorescent ribbon suspended from float to indicate subsurface currents that could cause problem sets	Results often inconsistent with behavior of net in water; concept abandoned
13. Set-position compass	Indicated optimum course for letting net go based on wind and current directions	No fleet acceptance; captains preoccupied by other events when sets are started; concept abandoned
14. Large volume purse seine	Based on purse-seine model studies, a net with increased depth-to-length ratio was constructed and tested	No alleviation of net collapse problems; testing suspended pending extensive modifications and repairs
15. Porpoise grabbers	Several designs of long-handled devices for moving porpoise into hand-release range	Some success with "shepherd's crook" design; success varies among individuals
16. Zipper system	Porpoise are isolated in apron section of net by gathering or pursing the base of the apron	Backdown unsuccessful when apron is "zipped-up"; poor separation of porpoise and tuna; concept abandoned
17. Black corks	Porpoise may be less reluctant to approach a dark corkline in the backdown area apex	Inconclusive results; paint wears off easily

III. Concepts to be tested in the future

18. Antitorque purseline	Torque balanced wire rope is less susceptible to rotating under loading which reduces incidence of rollups.	Some rollups continue to occur, but concept is promising; can be utilized in conjunction with antirollup purse block (see No. 19 below)
19. Antirollup purse block	Purse block designed to equilize forces on the two sides of the purseline to prevent twisting, thus reducing rollup problems	Prototype to be tested in conjunction with antitorque cable (see No. 18 above); concept considered promising
20. Guillen snap link	Speed-up set by allowing pursing and net hauling to procede simultaneously; prevents stern bend formation.	Concept considered promising
21. Da Rosa effect	Removal of section of floats which causes "scalloping" of net allowing pre-backdown escape of porpoise.	Concept based on observations made by one captain; may prove to be of some value
22. Polyvinyl panels	When used in apex of apron would provide a smoother surface for porpoise to escape over	Concept may prove to be of some value; further tests under fishing conditions are needed
23. Herding porpoise out of the net	Possible herding devices include the rescue raft, towed sections of corkline, or movable bubble curtains	Concept based on the fact that porpoises respond to changes in the location of disturbances; considered promising

out research directed specifically at reduction of porpoise mortality and improving stock assessment techniques. This dedicated vessel program will make possible a wide variety of research programs which could not otherwise be undertaken.

In spite of funding and vessel time problems, substantial progress has been made toward developing techniques for releasing porpoise from purse seines. Many of the concepts that have been explored are summarized in Table 11. In the first part of this table several concepts are described that have proven to be valuable (or at least very promising) in reducing porpoise mortality. Two of these, the backing down technique and the Medina safety panel, have already been discussed, and the rest will be considered next. The second part of Table 11 deals with concepts which have been tested to some extent and appear to be either ineffective or impractical; the third part treats concepts which may prove effective but which will require testing. Concepts falling into these two categories will not be described further in the text.

Even though backing down and the incorporation of safety panels in purse seines clearly reduced porpoise mortality, significant numbers were still being killed. At least three factors contributed to the continued mortality:

1. A certain number of porpoise were still becoming entangled in the net, even in the 2-inch mesh of the safety panel.
2. Canopies of loose webbing often developed along the side walls of the backdown channel in which porpoise could become entrapped and die.
3. A deep pocket would form at the release end of the backdown area with the net rising almost vertically to the corkline; when tuna surfaced in this area porpoise could be driven away from the release area and the backdown effort greatly inhibited for fear of losing fish.

To attack these problems of entanglement, canopy formation, and tuna surfacing in the apex of the backdown area, the NMFS, working with cooperating captains, undertook a series of experiments aimed at evaluating various modifications of the seine. These modifications involved adding a porpoise release apron to the net and use of finer-mesh webbing.

A porpoise release apron is simply a section of webbing that is added to the purse seine between the top to the safety panel and the corkline. Initially a trapezoidal design was adopted using 2-inch-mesh webbing. The wide base of the apron was attached to the top of the safety panel and its tapered sides and narrow top were attached to the corkline. This prototype apron formed a shallow shelf at the apex of the backdown area, and the weight of the porpoise themselves caused the corkline to sink, allowing their release. Also, the shoaling of the apron area tended to prevent tuna from surfacing among the porpoise, and when tuna did surface they tended to drive porpoise onto the apron and over the corkline. With porpoise plugging the escape area, a smoother, more continuous backdown was possible, rather than the conventional stop-and-go procedure. Furthermore, the gradual taper of the apron along the uppermost sides of the backdown area caused increasing tension toward the release end of the seine and helped to eliminate some of the slack webbing near the surface. This reduced the tendency for canopy formation.

While the addition of an apron seemed promising in terms of resolving the problems of canopy formation and of tuna surfacing in the apex of the backdown area, the problem of entanglement in the 2-inch mesh of the safety panel and apron remained. Hence in 1975 experiments were begun to test the obvious approach of using even finer-mesh webbing in the safety panel and apron. These tests showed that porpoise mortality owing to entanglement could be virtually eliminated in sections of the net where 1¼-inch-mesh webbing was used.

As testing of the modified purse seines continued, some problems were encountered. For example, compared to the normal 2-inch-mesh safety panel, the smaller 1¼-inch mesh when used in one type of safety panel and apron complex that was tested caused such an increase in drag that the corkline along the sides of the apron submerged during the early stages of backdown. This increased the danger of fish escaping and necessitated a slower backdown. In other instances, certain vessels were unable to submerge the corkline in the release area during the later stages of backdown, which increased the need for hand rescue.

To overcome problems such as these, further modifications of the safety panel and apron complex were necessary. These experiments culminated in development of a safety panel and apron configuration that, beginning in mid-1978, will be mandatory for all United States vessels licensed to fish on porpoise under the MMPA (Fig. 11; see also Fig. 5). This configuration features a triangular apron known as a "super apron." It also incorporates a double depth (i.e., two approxi-

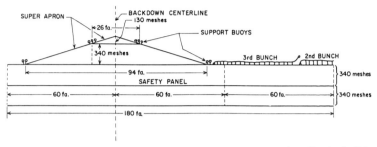

Figure 11. Design of the super apron system tested by the *Elizabeth C.J.*
(Source: NMFS; vertical distances not drawn to scale)

mately 5½-fathom-deep strips), 180-fathom-long safety panel which is much larger than the heretofore conventional safety panel (single depth, 120 fathoms long). Webbing of 1¼-inch mesh is utilized throughout the safety panel and apron. The super apron system was first tested in late 1976 aboard the chartered United States seiner *Elizabeth C.J.,* which was fishing under a special IATTC 1,000-ton yellowfin allocation for gear research. It was found that the super apron modification resulted in a more even downward pull on the corkline in the early stages of backdown, which eliminated the problem of corkline submergence. Furthermore, no difficulty was encountered in sinking the apex of the super apron. With the super apron system, the apex of the backdown area is more ramplike than shelflike, becoming progressively more shallow from the bottom of the seine to the corkline at the surface. This seems to be a better configuration in that porpoise have a greater tendency to remain separated from the tuna. Canopy formation is also minimized.

In addition to designing the purse seine so as to maximize the effectiveness of the backing-down procedure, other measures have been developed to reduce porpoise mortality. These involve the effective deployment and utilization of speedboats during the set and the use of a porpoise rescue raft. Because these procedures have proven to be of value, United States regulations have been established under the MMPA requiring that speedboats and rescue rafts be utilized in appropriate circumstances.

After the tuna and porpoise are encircled in a purse-seine set, speedboats can be used in several ways to help reduce porpoise mortality. One important technique is towing on the net to prevent its collapse from adverse wind or current conditions or from gear malfunction that causes a delay in the set. In a disastrous net collapse, the open surface area within a section of the net becomes so restricted that porpoise are forced into contact with the net, where they can become entangled or prevented from reaching the surface to breathe. (Historically, such disaster sets have contributed very substantially to the total porpoise kill.) In most instances speedboats can prevent net collapse disasters by towing on either the bunchlines or the corkline. Towing is a preventive technique rather than a remedial one. If a net has collapsed, it is generally not possible to reopen it by towing with speedboats. Hence, it is very important to be alert for conditions associated with impending net collapse so that properly deployed speedboats can take prompt preventive action.

Speedboats are also used to repel porpoise from areas where hazards may be present. In particular, they frequently circle near the stern of the vessel to keep porpoise out of the area where the net is being hauled aboard. During backdown one or more speedboats are normally stationed just outside the net near the apex of the backdown area with their motors off. There they can assist individual porpoise out of the net and also prevent escape of tuna by warning the captain and pulling up on the corkline if tuna approach. Hand rescue along the sides of the net can continue if necessary while the net is being hauled after backdown. Speedboats are also used to adjust the backdown area corkline to its proper configuration prior to initiation of backdown, particularly when seines are equipped with super aprons.

To take the concept of porpoise rescue one step further, several United States captains developed the idea of using a person in a small inflatable rubber raft in the backdown release area. This concept has been tested and improved by NMFS scientists on research charters. Equipped with a mask and snorkel, the raft man can signal when the area is clear of fish so that backdown can proceed. During backdown the raft can be used to herd porpoise toward the release area and drive them over the sunken corkline. The raft man can also signal to continue backdown until all porpoise are out of the net. This is especially important with respect to a behavioral pattern exhibited by spotted porpoise, which at times rest passively on the net in the bottom of the seine. In the past such animals have sometimes been assumed to be dead and backdown has been terminated. In fact, if backdown is continued, these porpoise can be released when they rise to the surface to breathe. It was also discovered that the raft man could hear vocalizations of porpoise that were still in the net even if they could not be seen, and underwater listening became the final check before termination of backdown on the 1976 *Elizabeth C.J.* cruise. In the final stages of backdown the raft man can also assist in hand releasing porpoise.

These efforts by the raft man and the speedboat operators to get the last few porpoise out of the net alive makes zero-mortality sets possible in cases when all porpoise cannot be successfully released by backing down alone. The significance of allowing a few additional porpoise to escape during backdown should not be overlooked. There are something like six thousand to seven thousand successful porpoise sets made annually in the eastern Pacific. If just a single additional porpoise could be released safely in each of these sets, this would represent a significant reduction in the total porpoise kill.

Adoption of the super apron and effective deployment of speedboats and a manned rescue raft together constitute an advanced fishing system that was thoroughly tested on the 1976 *Elizabeth C.J.* charter. Figure 5, which depicts a school of porpoise escaping over the corkline during backdown, gives a good idea of this fishing system in action. Notice the configuration in the water of the double-depth safety panel and the attached super apron. A speedboat is stationed in the release area to assist porpoise and to prevent escape of tuna. A raft man herds the porpoise toward the release area and checks to see that they all escape safely.

The results achieved on the *Elizabeth C.J.* cruise demonstrated that porpoise kill rates can be reduced to extremely low levels, at least under the conditions she encountered, which were optimal in all respects from a porpoise-saving point of view. In twenty-six normal porpoise sets (that is, sets without any gear malfunction or net collapse), 492.5 tons of yellowfin tuna were captured, and an estimated 19,512 porpoise were encircled with only 4 mortalities. Thus, the kill rate was only 0.0081 deaths per ton of yellowfin captured on porpoise or 0.15 deaths per porpoise set. Should this kill rate ever actually be approached throughout the international fleet, the total kill would be reduced to an extremely low level that most would consider acceptable. (In fact, reduction of the kill rate to even ten times this level would represent a very substantial achievement in reducing porpoise mortality.) Realistically, however, it is virtually impossible that the fleet as a whole can match the *Elizabeth C.J.* kill level, at least in the near future. First, even with diligent effort it is probably impossible for some vessels and crews to operate as effectively as the *Elizabeth C.J.* and its crew. For example, vessels without bow thrusters may not be as precisely maneuverable during backdown as the *Elizabeth C.J.* Second, in determining the *Elizabeth C.J.* kill rate, no allowance is made for disaster sets in which something goes seriously wrong that results in large porpoise kills. In the past a significant share of the total porpoise mortality has occurred in disaster sets, which will certainly continue to occur in the future, although hopefully with reduced frequency. Third, porpoise in the area fished by the *Elizabeth C.J.* offshore of Mexico and Central America have been subject to capture in purse seines since the early 1960s; normally they remain relatively calm in the nets, which facilitates their release during backdown. Further offshore, especially west of the CYRA where fishing did not start until the late 1960s, porpoise are more naïve, which may contribute to the mortality problem. Fourth, weather conditions were generally good during the *Elizabeth C.J.* cruise, and this will certainly not always be the case.

Nevertheless, the results of the *Elizabeth C.J.* cruise together with the significant reduction in the United States kill rate in 1977 both suggest that opportunities definitely exist for further reduction of the porpoise kill if advanced mortality-prevention techniques are adopted throughout the international fleet. Also, there is real hope that continuing research will lead to further breakthroughs in porpoise-saving technology.

In future research in this area, it is likely that behavioral aspects will play an important role as they have in the past. Studies on the behavior of tuna and porpoise in the eastern Pacific have a long history. Virtually all of the earliest work was carried out by the fishermen themselves. Although they probably did not think of themselves as being engaged in behavioral research, it was their behavioral observations that led to the development of the fishing gear and techniques that made a tuna fishery possible in the eastern Pacific.

The development of the original bait fishery was based on the behavioral observation that chumming with live baitfish could induce tuna to feed and thus be captured by pole and line jigging. Similarly, the observation that tuna often congregate beneath logs or other floating objects played an important role in the development of the purse-seine fishery.

Behavioral observations have also been basic in the development of purse-seining techniques for yellowfin associated with porpoise. The critical observation was that the bond between schools of tuna and porpoise (whatever its nature may be) was strong enough so that, if the porpoise school could be encircled by the purse seine, then the associated tuna could also be captured, even after a lengthy chase. From the point of encirclement on through to completion of the set, any porpoise mortality only hindered the fishing operation. For this reason it was obviously desirable to find methods for releasing porpoise from the nets unharmed, and once again behavioral observations played an important role. The fact that the porpoise and tuna schools generally occupy different areas within the net (the porpoise near the surface and away from the boat, the tuna deeper and often approaching the boat) makes possible the backdown maneuver. Similarly, the fact that porpoise will allow themselves to come into contact with the net at the risk of entangling themselves led to the original adoption of the 2-inch-mesh Medina safely panel, as well as to subsequent modifications using finer 1¼-inch-mesh webbing. Most recently, recognition of the spotted porpoise behavior pattern of sometimes resting passively on the netting in the bottom of the seine should lead to effective release of these individuals by continuation of backdown.

Internationalization of the Tuna-Porpoise Problem

Although the United States has taken the lead in addressing the problem of porpoise mortality during purse-seining operations, any real resolution of the problem will require the cooperation of all coastal nations that border on the eastern Pacific as well as all nations whose flag vessels harvest tuna in the area. The tuna-porpoise problem is truly an international one for several reasons.

First, porpoise, like the tuna that they associate with, are migratory and widely distributed and can rightly be considered an international resource. The populations of the major species, spotted porpoise and spinner porpoise, are believed to be made up of a number of races with individuals generally remaining within an average radius of about 600 miles. Thus, like the tuna, porpoise are capable of moving from waters within the juridical zone of one nation to those of another, or into high seas waters beyond national juridical zones. Clearly, if a management program is to be implemented, it should apply wherever the porpoise occur, and this implies cooperation among nations.

A second obvious reason for international cooperation is that the number of nations fishing in the eastern Pacific has increased steadily over the years. Furthermore, these new entrant fleets are in general made up of large, modern purse seiners capable of fishing throughout the entire region. Because the problem of porpoise mortality is common to the vessels of all participant nations, no fully satisfactory resolution can be achieved through the actions of a single nation. Instead all participants will have to cooperate.

There is another problem associated with addressing the porpoise mortality problem on a unilateral basis. If only one nation enforces strict regulations for protection of porpoise, then this could provide at least a partial incentive for vessels of that nation to transfer to flags of other nations without such regulations. If this were to happen on a large scale, strict regulations to protect porpoise could actually ultimately have a negative impact on the stocks.

The United States, recognizing these problems, has provided for import sanctions against other nations that do not act to protect porpoise. The MMPA states: "It is unlawful to import into the United States any . . . fish, whether fresh, frozen, or otherwise prepared, if such fish was caught in a manner . . . proscribed for persons subject to the jurisdiction of the United States." In applying this provision of the MMPA, it has been established that tuna can be imported into the United States provided that the director of the NMFS makes a finding that the non–United States fishing, although not in conformity with the specific requirements of United States regulations, was accomplished in a manner which does not result in an incidental mortality and serious injury rate in excess of that which would have resulted from fishing under United States regulations.

With the above interpretation, the import sanctions of the MMPA probably do provide a significant incentive for non–United States vessels to adopt gear and fishing techniques comparable to those required of United States vessels. Because of the nature of the world tuna industry, however, it will be difficult for any import sanctions to be completely effective. Of the roughly 1.6 million tons of tuna captured annually in the world, only about 7 percent is taken in association with porpoise. The world catch moves through many ports aboard both harvesting vessels and cargo vessels in both raw and processed form. To identify the origin of each cargo as well as the manner in which the fish were caught would be nearly impossible, especially if merchants chose to disguise this information. Furthermore, even if yellowfin caught in association with porpoise could somehow be accurately identified, such fish could be sold on a non–United States market (e.g., Japan or western Europe). Thus, a cooperative agreement among the nations involved would be preferable to unilateral United States action.

Fortunately, considerable progress toward international cooperation to resolve the tuna-porpoise problem has already been made within the context of the IATTC. The first step was to consider the rationale for commission involvement. In addressing this matter, the basic question was whether the commission, because of its international nature, could execute or coordinate any of the required research and management activities more effectively than nations acting individually. If so, a reasonable rationale would exist for commission involvement. After considerable deliberation, it was concluded that a legitimate rationale did exist for IATTC involvement in the tuna-porpoise problem, and at its annual meeting in 1976 three basic objectives were adopted:

1. To maintain tuna production at a high level
2. To maintain stocks of porpoise at or above levels that would insure their survival in perpetuity
3. To make every reasonable effort to ensure that porpoise are not needlessly or carelessly killed in fishing operations

The next step was to develop a program of activities pertinent to resolution of the tuna-porpoise problem. This was done at a special meeting of the IATTC held in mid-1977. In formulating the program, a basic consideration was to focus commission activities on areas where, because of its unique international character, it could accomplish its goals more expeditiously than could member nations working independently. Conversely, heavy involvement was avoided in research areas where national programs could proceed effectively on their own.

The single most important area of commission involvement would be to organize and operate an international seagoing technician program in order to facilitate estimation of total porpoise kill and porpoise population sizes. The commission would also keep abreast of all current gear and behavior research, organize gear workshops, and undertake computer simulation studies of the interacting tuna and porpoise populations. When appropriate the commission would make management recommendations to member governments designed to reduce the incidental mortality of porpoise. Proposed regulations could take a variety of forms such as regulations pertaining to gear and fishing techniques, kill quotas, time-area closures, and so on.

While considerable research has been done on porpoise in the past, especially in the areas of population estimation and stock identification, much basic research remains to be done. For example, the nature of the bond between tuna and porpoise in the eastern Pacific is not presently understood although a number of hypotheses can be suggested to explain the bond (e.g., porpoise and/or tuna may derive some advantage in feeding when they are associated, susceptibility to predation may be reduced, etc.). There is no particular basis for saying any single hypothesis is any more realistic than the others. However, since several hundred thousand tons of porpoise and perhaps roughly the same tonnage of tuna are associated in the same area it is reasonable to assume that there must be some strong interactive factor operating between them. Depending on the nature of these interactive factors, differential removal of one member of the species complex could have a profound influence on the abundance and distribution of the other members. Let us assume, for example, that tuna and porpoise compete for food and are food limited, but associate together for some other reason. If the number of porpoise is increased by protecting them, then the standing stock and potential yield of tuna might be reduced. Likewise, reduction of the tuna stocks might cause the number of porpoise to increase. In fact, one might speculate that porpoise stocks may have increased considerably over the period of years that large catches of tuna have been taken in the eastern Pacific. It is also possible that the tuna stocks and their potential yield may have increased with the advent of porpoise fishing and consequent porpoise mortality.

Very little research has focused on the porpoise-tuna bond, although some sampling of stomach contents of tuna and porpoise from the same schools has been done by the NMFS. Such food habit studies are a straightforward approach to seeking an understanding of the bond. But even these studies require extensive sampling of the stomach contents of both porpoise and tuna from the same schools, as well as data on what they are eating relative to what is available in the ocean, and how the composition of their diets changes relative to changes in the availability of food organisms. Such studies will obviously take a number of years with the probability of success being uncertain.

In addition to contributing to an understanding of the relationship between porpoise and tuna, studies of porpoise behavior are also important from the viewpoint of the tuna fishing process. For example, why do porpoise form schools and why are some schools large and others small? Do the larger schools have more tuna associated with them? What happens to a porpoise school after it is set on? Do the members scatter and reform a new school and, if so, do they tend to avoid tuna? Do porpoise learn to avoid tuna boats or areas where tuna boats operate? If so, what effect does this have on tuna fishing success? If such learning occurs, is it developed enough to insure perpetuation of porpoise stocks? Will tuna catches decline as learned behavior in porpoise increases? If porpoise population estimates are based on encounters with vessels, and if porpoise learn to avoid vessels or areas where they operate, how will this affect population estimates? Does stress associated with fishing affect the physiology of the animals or cause changes in mating, feeding, and parental behavior? Why are more females and young porpoise apparently being caught and killed than males? What happens to tuna fishing success if sex ratios are altered? Can porpoise behavior be utilized to ensure their protection while at the same time sustaining the tuna harvest? Why do porpoise associate with tuna only in certain subareas of the eastern Pacific and there only with larger yellowfin rather than smaller yellowfin or skipjack? Although tuna and porpoise exist together in other ocean areas, why do they commonly form school complexes only in the eastern Pacific? While some information has recently become available, it is obvious that much further research is needed if these and similar questions are to be fully answered.

While an international body can play an important role in coordinating and expediting research in many of the areas mentioned above, the major share of the work will undoubtedly be carried out by public and private research organizations in various nations, including both the RANs and the non-RANs that participate in the fishery. Much of the at-sea work that will be essential to the success of these studies will have to be carried out within the 200-mile zones of coastal states from aboard both research and fishing vessels. The obvious need to facilitate such studies points up, once again, the need for international cooperation in resolving the porpoise mortality problem.

Billfish Conflicts Involving Commercial and Sport Fishermen

When one thinks of deep-sea sport fishing one thinks of a tackle-busting marlin crashing through the surface of a tropical sea. This is a familiar picture from the decks of sport fishing boats in numerous resorts around the world. But besides providing such thrilling sport, the marlin and other billfish support major commercial fisheries throughout the tropical and temperate seas of the world. Like the tuna which they are caught in conjunction with, billfish are highly migratory, and their proper conservation and management requires international cooperation. Unlike the tuna (with the exception of Atlantic bluefin off the United States and Canada) billfish are sought by two different principal users—sport and commercial fishermen—with increasingly important conflicts resulting. To resolve these conflicts and develop a rational program of conservation and management will require a good understanding of the biology and behavior of the animals themselves, and the perceived needs of the major user groups will have to be carefully considered in order to achieve a series of well-thought-out compromises.

From both behavioral and taxonomic points of view billfish are very similar to tuna. They are caught in the same areas, at the same times, and in many cases with the same gear as tuna. They are generally included as tuna-like fishes in world catch statistics compiled by the Food and Agriculture Organization of the United Nations as well as by most national and international organizations. Taxonomically the billfish are included in two families (Istiophoridae and Xiphiidae) in the suborder Scombroidei, which also includes the family Scombridae. The tuna are within this last family.

There are nine species in the family Istiophoridae, the most familiar of them being the sailfish *(Istiophorus platyperus),* the striped marlin *(Tetrapturus audax),* the white marlin *(T. albidus),* the blue marlin *(Makaira nigricans),* and the black marlin *(M. indica).* The second family, Xiphiidae, contains only one species, the swordfish *(Xiphias gladius).*

Of the ten species of billfish the sailfish, the blue marlin, and the swordfish have the widest distribution, being found throughout most of the tropical and subtropical oceans of the world. The largest concentrations of sailfish seem to most often occur within a thousand miles of land, whereas swordfish and blue marlin appear to be as abundantly distributed in waters far from shore as they are in near-shore waters. Although only limited tagging of these species has been carried out, what has been done demonstrates extensive migrations (Fig. 12). For example, white marlin tagged off the east coast of the United States have been recovered off Brazil, Venezuela, and Mexico.

Striped marlin tagged off the west coast of Mexico have migrated as far as 4,000 miles to the west and over 2,000 miles to the south. Black marlin tagged off northeastern Australia have also demonstrated long-range migrations to New Zealand, northern New Guinea, and to Nauru west of the Gilbert Islands. Sailfish are also highly migratory. While their migrations are not shown in Figure 12, individuals tagged off the east coast of the United States, like white marlin, have been recovered off Brazil, while others tagged off the Yucatan Peninsula in Mexico have migrated to Venezuelan and eastern United States waters. Judging from these tagging results, billfish appear to be just as highly migratory as their close taxonomic relatives, the tuna.

Behaviorally billfish are quite different from the tuna in that they do not congregate in schools; rather they tend to maintain a distance between themselves and their nearest neighbor. Thus, the distribution of individuals appears to be more analogous to that of territorial animals than to that of the generally gregarious tuna. Because of this behavioral characteristic, billfish are not harvested by bait fishing or purse seining, the normal surface fishing methods used for tuna. Instead, they are harvested almost exclusively with longline gear. Longline gear does not fish selectively as to the species it captures. It would not be unusual for a single set of gear to capture as many as three species of tuna and three species of billfish. However, by concentrating effort in certain geographic areas, the species composition of the catch can be altered, but not too greatly.

In Table 12 catches of billfish are shown for 1975 by major species groups and by ocean areas. Although billfish make up only a small portion of the total catch of the tuna-like fishes (3–4 percent), in certain areas they represent a significant share of the longline catch. For example, in the eastern Pacific Ocean they are very important in the longline catch, particularly in the near-shore areas off Mexico, where they are the mainstay of the longline fishery.

Prior to the expansion of the longline fishery after World War II, only swordfish were taken in large commercial quantities. With the spread of longline fishing and the increased catches of billfish that resulted, a market developed in Japan for marlin, sailfish, and spearfish for use in the production of fish sausages, fish hams, and sashimi. On the other hand, much of the swordfish taken by longline gear has been exported to North America and western Europe for the fresh fish market.

The billfish fishery is economically important, with the dockside

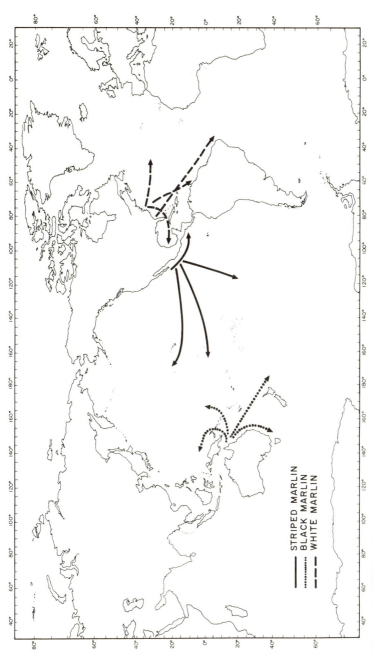

STRIPED MARLIN
BLACK MARLIN
WHITE MARLIN

Figure 12. Billfish migrations as determined through tagging programs undertaken by various sportsmen groups and agencies

TABLE 12

BILLFISH CATCHES BY COMMERCIAL FISHING OPERATIONS IN 1975 BY
ALL NATIONS COMBINED
(Thousands of metric tons)

Species group	Pacific Ocean	Atlantic Ocean	Indian Ocean	Total
Blue and black marlin	15.8	1.3	4.2	21.3
Striped marlin	18.0	–	1.4	19.4
White marlin	–	1.1	–	1.1
Sailfish and spearfish	7.6	0.6	1.7	9.9
Swordfish	15.4	11.8	1.6	28.8
Unspecified billfish	0.5	2.1	0.6	3.2
Total	57.3	16.9	9.5	83.7

value of the catch being close to $100,000,000 annually. When retail price is considered as well as the value of ancillary industries, its value is much greater. Demand for the product is high and increased production is sought. However, recent analysis concerning the impact of exploitation on the abundance of billfish stocks suggests that increased fishing effort will not result in increased production. Indeed, increased exploitation could well result in reduced production.

Billfish catches by sport fishing are not nearly so well documented as they are for commercial fishing; in fact they are hardly documented at all on a world scale. There are a number of major big-game fishing centers (e.g., Piñas Bay, Panama; Cairns, Australia; Cabo Blanco, Peru; Acapulco, Mexico; Kona, Hawaii; and Bermuda), but even in these major centers no system for comprehensive data collection is routinely operated. In addition, there are literally thousands of ports from which tens of thousands of small, private sport fishing boats operate. To date it has been virtually impossible to estimate the number of boats and people involved in sport fishing for billfish, the amount of time that each boat operates in pursuit of billfish, or, most importantly, how many billfish of each species are caught. It is clear, however, that a large number of people spend great sums of money in pursuit of billfish for sport. Like the communities that have been built around commercial fishing operations in various parts of the world, many communities have been built up as sport fishing centers. Such centers are generally not exclusively devoted to sport fishing, however. Other tourist-associated pursuits are important to the economy of such centers because areas where billfish occur in abundance are generally areas where tropical beaches and a host of associated activities such as swimming, boating, water skiing, and sunbathing are available. It is virtually impossible to objectively partition the expenditure of the tourist recreational dollars among the various recreations. But it is logical to assume that big-game fishing in such centers increases the options open to a tourist who is planning to spend vacation time and money. The more recreational options available for tourists in an area, the more attractive that area is to tourists. For that reason alone, big-game fishing has a significant impact on the tourist economy in many parts of the world, and this must be given proper consideration in the management of the ocean's renewable living resources.

Conflicts between sport and commercial fishermen have existed for a long time. In the past these have generally been confined to waters within a few miles of shore. Because of the developing nature of big-game fishing, however, these conflicts have intensified and entered the oceanic realm. Big-game fishermen now often fish in areas one hundred miles or more from shore, side by side with commercial vessels that heretofore had the "briny deep" to themselves.

Although the objectives of commercial and sport fishermen coincide in that they both want to catch fish, they differ remarkably in the way that they assess the quality of their fishing. In most cases the primary objective of commercial fishing is to make a profit on investment. This has often been considered to be consistent with maximizing the level of production that is sustained, even though economic theory suggests that this is not the case. But regardless of whether the management is maximum sustainable yield or maximum economic yield, the optimum population size that will sustain such a yield from a stock of fast growing fish like tuna or billfish is substantially less than the virgin population size. Also, the average size of fish in a population in equilibrium at its optimum size is much smaller than the average size of fish in an unexploited or virgin population.

In contrast, big-game fishermen are interested in catching large fish on the average and also in having a high probability of catching fish. Both these objectives are maximized when a population is at its maximum size, that is to say, in its virgin state. It is obvious that there is a fundamental conflict between these objectives and the objectives of commercial fishermen. If commercial fishermen are to sustain high levels of production from the billfish populations, both the average standing stocks of billfish and the average size of individuals must be reduced. This means that the average size of the fish taken by sportsmen goes down as well as the number of fish caught per unit of fishing time. If sportsmen are to catch greater numbers of larger fish, then commercial fishing on billfish populations that also support sport fisheries must be limited, at least to some extent.

Conflicts between sport and commercial fishermen over management objectives are already of significant importance, especially in view of the general trend for nations to extend their jurisdiction seaward to 200 miles. A notable example involves Japanese longline fishing off Mexico. Mexico has for some time had national legislation prohibiting the commercial exploitation of billfish within its juridical zone. Japan takes large catches of billfish and tuna in their longline fishery in the eastern Pacific Ocean, with the highest catches of billfish being made close to the Mexican coast. When Mexico's juridical zone extended to 12 miles, the fact that Japanese vessels had to fish beyond 12 miles did not pose a problem to the Japanese fishing industry. However, since extending their jurisdiction over fisheries to 200 miles, Mexico has prohibited the Japanese from fishing for billfish within that zone. Because longline gear is nonselective with regard to the species caught, billfish prohibition means prohibition of longline fishing, and this is what Mexico has done. The economic effect of this longline prohibition has been devastating to the Japanese industry. Not only have they been excluded from a lucrative billfish fishery, but their catch of other tuna species, particularly the bigeye tuna, has also been reduced substantially.

Similar unilateral action is being considered in the United States. In that country domestic law requires that all highly migratory species be managed by international organizations both within and beyond the 200-mile exclusive economic zone rather than managed unilaterally. For the purposes of this domestic law, tuna are considered to be highly migratory, but billfish for some reason are not, so they are subject to unilateral management. Considerable efforts have been made to prohibit the capture of billfish by longliners within the United States 200-mile zone. Most of this pressure is from sport fishing interests who fear that such longlining has severely damaged sport fishing for marlin. If longlining is proscribed within the United States 200-mile zone, the Japanese will suffer further economic hardship, particularly around the Hawaiian Islands, in the Gulf of Mexico, and off the east coast states.

In view of this dilemma of conflicting objectives, the question arises as to whether the interests of sport fishermen or the interests of commercial fishermen should be recognized, and if so, whether this should be to the total exclusion of the other's interests. The answer to this question seems obvious from both a moral and a social viewpoint: The interests of both user groups should be recognized in developing a rational management system for billfish resources. To accomplish this goal it is particularly important that dialogue between the two groups be fostered and exchanged.

Regardless of whether or not billfish are legally categorized as highly migratory species, they are in fact very migratory as was shown earlier. To be effective the management of billfish, like the management of tuna, must apply to the entire stocks of animals being considered. Management by unilateral coastal state action will not be effective for most billfish species, perhaps not for any of them. Therefore, billfish management must be carried out through international cooperation of the major states participating or having an interest in their fisheries. This includes sport fisheries as well as commercial fisheries. It does not appear necessary to choose one type of fishing over the other. There are various management options open. These might include, among others, the establishment of sport fishing sanctuaries in areas of concentrated sport fishing effort, quotas on the catch made by both types of fishermen, and open and closed seasons and areas.

Before realistic management schemes that are responsive to the needs of the interested parties can be formulated, a much broader base of information on billfish is needed. Good catch and effort statistics must be available for both types of fisheries, not just the major commercial fisheries. A better understanding of basic biological parameters such as those describing growth and mortality are needed. Also, studies of population structure and distribution are required. This all adds up to a major scientific effort directed toward the study of billfish on a broad geographic scale. Only through the acquisition of such data and information can effective billfish management programs be developed that include resolution of disputes among competing user groups.

13   Institutional Arrangements for Management of Tuna and Associated Species

Four general problems of tuna management—the collection and analysis of data for management of the fisheries, the distribution of catches among harvesters, economics and fleet carrying capacity, and the enforcement of conservation regulations—have been discussed in great detail in preceding chapters. A number of possible management systems were described and evaluated from this perspective because management, if it is to be successful over the long term, must deal with all of these problems effectively. In addition, management problems pertaining to porpoise, billfish, and other species captured in association with tuna must be dealt with. None of the currently existing organizations for tuna research and management is dealing effectively with all of these problems, either through lack of coordinated initiative or through lack of sufficient authority. Insufficient legal authority does not seem an insurmountable obstacle, however, because the conventions establishing existing international bodies are written in rather broad language and are open to wide interpretation. The real obstacle is the absence of political determination in concerned nations to recognize these problems and to set about dealing with them effectively. Once the political determination to resolve management problems is instilled, the details of the solutions and the institutional arrangements to implement them should be straightforward.

Need for Coordinated Global Management

In structuring a management organization the unique characteristics of the tuna and associated species and the fisheries that exploit them must be kept in mind. These have already been discussed, but are worth reiterating. These species undertake extensive migrations throughout their lives, some species much more so than others. They do not recognize man-made boundaries, nor are their natural geographic distributions compatible with any set of boundaries that might be established by nations in the foreseeable future, for example 200-mile juridical zones. If management is to be applied successfully on such stocks, that management must apply over the entire stock range. For a species such as southern bluefin this implies management responsibility applying throughout all the temperate oceans of the southern hemisphere. For other species such as albacore and northern bluefin, management must apply across entire oceans. Management applying to only a portion of a stock will not be effective over the long run.

The fleets that capture these migratory species are even more mobile than the animals themselves, with vessels capable of fishing in as many as three oceans in a single year. This fleet mobility makes possible significant alterations in fishing strategy in response to regional management actions. If a closed season or area is established in one region, the likely result will be for a share of this highly mobile fleet to shift its effort during the closed season to another area or ocean. Many of the world's tuna fisheries already have more vessel capacity than is needed to harvest the available catch and indeed some are in need of regulation. Therefore, such transfers can only exacerbate problems in these areas. Fleet mobility also has important implications from the viewpoint of establishing efficient data systems because basic data used in studying population dynamics are collected from the vessels themselves (catch and effort statistics and length frequency samples).

The present trend toward structuring international institutions for the management of tuna and associated species on a purely regional basis seems unsound in that it fails to take into account the mobility of the fish themselves and the vessels that fish for them. It would be more reasonable and efficient to broaden the geographic scope of organizations responsible for research and management of tuna and associated species. In fact, a single management organization for these species operating on a global basis would be the most effective arrangement of all in terms of gathering statistical information and coordinating regional management decision making.

Many vessels travel from the zone of responsibility of one regional organization to that of another. It is obviously redundant for each organization to maintain its own separate data bank, group of technicians, and so on, in order to monitor these vessels as they move from area to area. It would be far more efficient to lodge responsibility for collecting catch and effort statistics and size composition information in one global body. Then a single cadre of technicians could collect the required statistical data from the fleets no matter where they operated. Also, a single depository could be maintained where data would be readily accessible for analysis, and data from various fisheries would be readily comparable.

A single global body could provide a focal point for coordination among nations in a number of other important respects. For example, it could insure that scientific research necessary to provide guidelines for management is carried out effectively and made available for each important stock. Furthermore, management recommendations formulated under the auspices of a global body could be designed to minimize effects of regulatory actions in one fishery on fisheries in other parts of the world resulting from fleet mobility. In dealing with the difficult problems of catch distribution, a global body should be in the best position to produce lasting solutions. Some national fleets fish extensively on two or three major fishing grounds during a year, while other less mobile national fleets are restricted to more regional patterns of exploitation. When shares of the potential harvest of tuna from a region are allocated, one important consideration should be whether any of the participant fleets fish in other areas. If they do, this should be taken into account in developing an allocation formula. Logically this could best be accomplished by a global body with broad responsibility, one not restricted to working only in a specific region. Similarly, the problem of excessive fleet capacity in many tuna fisheries could be effectively dealt with by a global body. One of the most promising approaches to control of fleet size lies in identifying areas with underdeveloped fisheries and arranging for gear transfers into them from overdeveloped fisheries elsewhere—a natural role for a global body. Finally, enforcement and surveillance activities could be most efficiently and economically carried out on a global basis.

If a single global body for the management of tuna and associated species is called for, how should it be structured so as to deal most effectively with the four major problem areas of catch determination, catch distribution, economics and fleet capacity, and enforcement? Several approaches to structuring a management body are possible, and some of these will now be discussed. Subjects to be covered include the various functions of a management body, the degree of separation between the different functions, staffing policies, membership criteria, voting methods, species coverage, and arrangements for regional decision making within a global body. While these matters will be discussed within the context of a single global management body, everything that is said would apply equally to regional management bodies with the exception of the last topic listed above.

Functions and Division of Responsibility

A global organization with responsibility for all matters related to management could be designed along the lines of existing regional bodies such as the IATTC and the ICCAT. A political entity with representation from all participating nations would be the governing body of the organization. This governing body would be responsible for adopting management policies and regulations that would then be submitted to member nations for implementation with respect to each nation's own flag vessels and nationals. It might also have enforcement responsibilities. As with present international commissions, the global organization could establish one or more secretariats or directorates to carry out its technical functions and, if it has any, its enforcement functions. The directorate(s) would provide technical advice to the governing body on matters relating to catch determination, catch distribution, economics, control of fleet capacity, and marine mammal conservation. There might be additional directorate responsibilities related to enforcement, although any enforcement activities should be carefully delineated from other directorate activities.

Directorate activities can be thought of as falling into either two or three basic categories, depending on how enforcement is handled. The first category would include all stock assessment and population dynamics studies necessary to provide technical advice to the governing body on potential yield levels from the various tuna fisheries of the world and the condition of related marine mammal stocks. This would entail collection of basic information from individual vessels such as measures of fishing effort expended, corresponding catches, and samples of the size composition of the catches; statistics on the international fleets including vessel characteristics; estimates of porpoise mortality and population size; and any other information necessary to estimate potential yields and to determine the effects of man's exploitation on stocks, including pertinent biological studies of a longer-term nature. Another area of responsibility would be research on gear and related matters for reduction of porpoise mortality. These activities would be carried out by an independently recruited staff of scientists and technicians who would work in conjunction with national fisheries agencies around the world. This staff could act as a coordinating body for related national research efforts.

In addition to fisheries scientists and technicians, professionals in the field of economics would have to be employed. Their activities would constitute the second category of directorate responsibility. Working with fisheries scientists, they would provide advice on the possible outcomes of different management approaches to the problems of catch allocation and control of fleet capacity. The economics branch could also have important bookkeeping responsibilities. For example, if a management system is implemented employing the participant fee concept, the economics branch could be responsible for collecting and disbursing funds.

Enforcement would constitute a third category of activity, but only on the assumption that the international management body is assigned responsibility for monitoring fishing positions of vessels of cooperating nations and inspecting their catches when landed to insure compliance with regulations. A satellite surveillance system or a radio triangulation system could be operated by the organization under the auspices of an enforcement branch. An international catch inspection team could also be established. Any violations of regulations would be reported through the governing body to the nations whose flag vessels were in violation for subsequent action. Under no circumstance would the organization itself impose penalties or levy fines. Member nations would remain responsible for detecting and apprehending vessels of noncooperating nations fishing illegally within their 200-mile zones.

Given these three categories of work, to what extent should activities in each category be kept separate from activities in other categories? Traditionally those aspects of management involving the technical methodology of fisheries science have been kept separate from activities in the economic, political, social, and enforcement realms. Fisheries science in this context involves the gathering, analysis, and interpretation of factual statistical and biological data in order to assess the effects of exploitation and the environment on the abundance and distribution of stocks. Collection of scientific data to carry forward stock assessment studies relies heavily on the willing cooperation of the fishing industry in making available their basic data to the scientists. To insure the flow of these data, fisheries scientists have always had the responsibility to maintain such records in the strictest of confidence, never revealing the activities of individual operators or companies; and in the great majority of cases the data have been protected from misuse by political entities for economic or enforcement purposes. The reasoning has always been that if data were misused for such purposes, then industry cooperation would be destroyed and the easy and efficient flow of data for scientific research ended.

Similar considerations would apply to a global management organization. Suppose a single directorate was organized into a scientific branch, an economics branch, and an enforcement branch with free flow of information and data among the three branches. Because data collected in connection with stock assessment activities by the scientific branch could be used in economic studies and for enforcement purposes, it is obvious that the scientific branch would risk serious loss of credibility with industry as well as with nations themselves, a situation hardly conducive to sound management decision making. In fact, intermingling of scientific, economic, and enforcement activities could jeopardize the very existence of such an international management organization.

The problem of separating functions is especially important in the case of enforcement. How then might enforcement be kept entirely separate from other management activities? Two approaches come readily to mind. One would be to separate the scientific and economic branches from the enforcement branch. In such a scheme two directorates could be established, one with responsibility for the scientific and the economic branches, and the second with responsibility for enforcement. The two directorates would have no line functions between them, but rather would report independently to the organization's governing body. In this way data collected for scientific and economic purposes, which must be kept confidential to insure credibility and a continued flow of unbiased information, could not be used in actions against vessels or individuals violating conservation regulations. An even stronger approach would be to delegate enforcement responsibilities to a body completely independent of the management organization. That is, a separate international agreement could be made among the tuna fishing nations to establish an independent body solely for enforcement of conservation regulations. As before, such a body would be responsible for monitoring fishing locations and unloadings for member governments. Violations would be reported to flag nations for appropriate action. A disadvantage of this approach would lie in the added costs of establishing and operating a second independent body, but advantages in terms of maintaining credibility and insuring a continued flow of data might offset these added costs. Problems concerning credibility and data flow could also arise from intermingling scientific research with activities of the economics branch. Certain types of data that are important in assessing stock condition would also be essential to economic studies aimed at evaluating possible schemes for catch distribution and limitation of fleet growth. For example, data on the geographic distribution of catches would play a key role in establishing catch allocations, and catch rate data would be basic to any program for control of fleet size. Because controls and limitations could well result from such economic studies, nations and industry entities might tend to bias or restrict the flow of essential data concerning the fishery. This could cripple the resource management program. To avoid this problem, consideration might be given to establishing separate directorates for scientific and economic activities. Complete isolation of the scientific branch from economic analysis and enforcement functions would encourage the highest possible level of cooperation by nations, harvesters, and processers in providing the data required for management of the resources.

Despite the arguments for separating scientific research from economic analysis and enforcement functions, any final decision on organizational structure will require careful consideration. It should be kept in mind that a single directorate could probably function more efficiently than multiple directorates in advising member nations on the wide range of matters relating to the fishery. Hence, the basic question is whether the quality of the scientific data necessary for making conservation decisions would deteriorate to an unacceptable degree as a result of combining functions under a single directorate. Another factor to consider is that much of the data on vessel activities collected by an economics branch or an enforcement branch could be very valuable in making conservation and management recommendations to member nations. This suggests the possibility of creating a single directorate organization with very scrupulously controlled (and well-publicized) one-way data flows from the economics and enforcement branches to the scientific branch. Such an approach would simplify the structure of an international tuna management organization and encourage efficient operation without compromising vitally important data sources.

Staffing

No matter how an international tuna management organization is structured and its functions separated or combined under directorates, the question of how it should be staffed should be carefully examined. The most fundamental purpose of such an organization is to conserve and manage the resources. To accomplish this, certain basic data must be collected and analyzed which requires the specialized skills of a trained cadre of scientists and technicians. There are differing views concerning the administrative framework under which this can best be accomplished, and these views are reflected in the existing commissions concerned with tuna conservation and management that were discussed in Chapter 3. On the one hand, there is the IATTC which has a full scientific staff that carries out all necessary research under the supervision of a director of investigations who serves at the pleasure of the commission. On the other hand, there are organizations such as the IPFC and the IOFC which have no permanent staff whatsoever except an elected chairman and vice-chairman who are fully employed at other jobs. While these organizations have undertaken such activities as sponsoring symposia, they have produced very little in terms of management research. Between these two extremes is the ICCAT which has a permanent secretariat to attend to certain fiscal and administrative matters, but which relies on panels and committees composed of scientists from national sections for data collection and analysis. To determine which of these approaches provides the best solution to the staffing problem, matters such as effectiveness, accountability, impartiality, responsiveness to the needs of member nations, and past experience must be considered.

First as a matter of general principle, if an organization is to be effective, responsibility for specific jobs must be determined and accountability established. In the case of a tuna management organization, collection of the basic data required for the scientific studies that underlie management decision making is of fundamental importance. For purely scientific purposes data need not be collected on a rapid basis, and lags in data acquisition of more than a year are acceptable. But for management purposes data must be collected on a real time basis. Catches on board vessels at sea must be closely monitored both inside and outside conservation and juridical zones in order to determine closure dates and modify catch quotas. Responsibility for collecting such data on a real time basis and with sufficient accuracy and precision must be lodged within a body that is fully accountable to the parent organization. In the IOFC and the IPFC no responsibility for data collection has been delegated and hence there is no accountability. With neither a full-time secretariat nor a staff, it is not surprising that basic statistics for stock assessment have been lacking in the Indo-Pacific area. In terms of data acquisition for stock assessment, the ICCAT has made substantial progress in recent years although data availability still restricts the extent of certain stock assessments. Also, data are still lacking for the kind of real time management decision making that would be required if overall catch quotas were to be established in the Atlantic. In the eastern Pacific, the IATTC staff has been able to collect data on the geographic and temporal distribution of catch and effort, the size composition of the catch, and numerous other factors which have facilitated the scientific studies required for conservation of the resource. The commission has also been able to set up a real time data acquisition system for management of the resource. The data collected on the fishery have provided the basis for many economic, social, and legal studies carried out by member governments.

It seems clear that if the challenges of tuna management and porpoise conservation are to be met in a timely and efficient manner, any newly formed organization must at the minimum be provided with sufficient fiscal support and legal authority to retain an independent and internationally recruited staff to collect and compile basic data on catch, effort, size composition, vessel characteristics, and porpoise populations. By assigning these responsibilities to a qualified technical staff within the organization, member nations can hold that technical staff accountable for getting its important tasks accomplished. The staff would also be responsible for maintaining the confidential nature of the data they gather, an absolute necessity.

Though scientific research should be, and generally strives to be, free from political and economic pressures, it is not always possible to avoid them entirely. This unfortunate fact can lead to questions concerning the impartiality of stock assessment studies carried out by nations with strong interests in the tuna fisheries. This problem becomes especially serious when developing nations that lack sufficient national expertise in fisheries science are confronted with findings by scientists of developed nations. Many developing nations, some falling within the latitudinal zones in which tuna occur, do not have a long history of technological development (in some cases, none whatsoever), and thus they lack adequate numbers of trained fisheries scientists. Even if there is a scientific nucleus, it may be critically needed for nonresearch tasks. For example, in many developing nations the most important concern is fisheries development, not stock assessment and fisheries management. In such countries administrators may not be very eager, and indeed may not deem it in the nation's best interest, to rely solely upon the scientific expertise of developed nations which are members of an international management organization. It seems natural for them to suppose that technologically and economically developed nations may be interested in research oriented toward objectives completely different from those of developing nations; and it is possible that research conducted by national sections of an international organization may reflect the views and needs of their own nations, especially if there is no opposing team of scientists to review their findings. Another danger for an organization without an independent research staff is that seemingly minor differences in research results presented by various national sections may be allowed to obscure what might be an adequate scientific basis for making needed management decisions. These problems can be largely avoided if an organization has its own internationally recruited scientific staff. Such a staff would serve all member nations equally, and its findings would be more likely to be accepted as being impartial than conclusions presented by national sections of nations competing for the resources.

The need to develop expertise in fisheries science is critical in many underdeveloped nations, and an international organization could and should be responsive to this need. Organizations without scientific staffs could institute a series of training courses, and such courses have, in fact, been organized by the ICCAT with success. They provide opportunities for technicians and scientists in developing nations to gain experience and training that might otherwise be difficult for them to obtain. An organization with its own scientific staff could also institute training courses, but it would be able to provide technical assistance in other ways as well. For example, the IATTC has received a large number of visiting scientists, mostly from developing member nations, for advanced training. Similarly, staff members from time to time have taught special short courses in fishery biology, population dynamics, and oceanography in member nations. The IATTC staff has also often provided technical assistance to member nations on matters other than those pertaining directly to tuna.

Development of expertise in fisheries science in underdeveloped nations is also stimulated directly through participation in the deliberations of an international management organization. In the ICCAT, for example, scientists from underdeveloped nations are directly involved in the work of scientific committees that make management recommendations to the parent body. In the case of the IATTC, national scientists evaluate the analysis and recommendations of the permanent scientific staff and serve as advisors to their delegations. An old saying has it that experience is the best teacher. With this in mind it is of interest that international fisheries commissions with independent research staffs have generally succeeded in carrying out the data collection and research necessary to recommend management measures for conservation of species under their jurisdiction, and that such conservation recommendations are in effect. In addition to the IATTC, these scientifically staffed commissions include the International Pacific Halibut Commission and the International Pacific Salmon Fisheries Commission. On the other hand, organizations without their own staffs have varied greatly in their degree of success, ranging from ineffective to, in the case of the North Pacific Fur Seal Commission, highly effective.

Considering all factors, it seems apparent that any international organization established for the scientific study, management, and conservation of tuna, porpoise, and related species would function most effectively if endowed with an independent scientific research staff. A similar case can be made for an independent staff to carry out the functions of an economics branch. Such a staff would be responsible for gathering and analyzing economic data in an objective and impartial fashion, for providing training and advisory services in economic areas to member nations, and for administering a participant fee system should such a system be established.

Membership and Voting

How inclusive or exclusive should criteria be for membership in a global organization for management of tuna and related species? This is an important issue because of the large number of nations that have an interest, great or small, in these resources. Normally this interest would focus on the commercially valuable species of tuna (and billfish as well), but an interest in the protection of porpoise could provide an alternate reason for seeking membership.

To recall certain pertinent facts discussed in Chapter 2, there are at least three reasons why a nation might be interested in the utilization and management of tuna resources. First, the various principal market species and secondary market species occur off the shores of nearly all nations bordering tropical and temperate waters. If only the principal market species are considered, commercial concentrations are found off the coasts of thirty to forty of these nations, and they all have an interest in tuna management based on their adjacency to the resources. Furthermore, their interest would take on added significance under an extension of fisheries jurisdiction to 200 miles on a worldwide basis. A second reason for interest in tuna management is participation in the fisheries themselves, either off a nation's own coast or on a more widely ranging basis. During 1974 about forty nations harvested tuna on a commercial basis, but their catches varied greatly in magnitude. Some of these nations took very small catches of less than 500 tons per year, while two nations, Japan and the United States, accounted for 55 percent of the world catch of principal market species, and just six nations took 77 percent. In addition to these participants, a number of other nations have demonstrated an interest in entering the fisheries, some of them non-RANs with respect to the fisheries that they wish to enter. A third basis for interest in tuna lies in utilization of the catch, where once again a few nations dominate. The United States and Japan consume about 75 percent of the catch, another 20 percent or so is utilized in a few western European nations, and the rest is dispersed in small quantities among other nations.

Two of these three interest areas would appear to provide a rational basis for determining membership in a global management organization—adjacency to and participation in the fisheries. An important question is: Should all of the nations that make significant catches or are adjacent to large resources be included? If all are included, the organization would obviously be very large indeed—perhaps too large to function effectively. But the dangers of excluding nations are also great. Exclusion of nations runs counter to the concept of encouraging coastal nations to allow open access to a regional resource within their 200-mile zones. Also, if a nation is excluded because it is not an important participant, and it later develops a significant fleet that operates independently of the international organization, this could jeopardize the management program. Thus nations, if excluded, clearly have the potential to disrupt any agreement significantly. This leads to the conclusion that instead of limiting membership, all participants in the fishery and all resource adjacent nations should be encouraged to join and become integrated into a cooperative international management program.

In establishing any international organization for management of tuna and related species, the decision on how voting is to take place is of vital importance. This is especially true for a membership as large as that contemplated in a global body, for the greater the membership the greater the difficulty in reaching agreement. There are a variety of voting systems that have been used in other international bodies, the most common ones being unanimity, simple majority, and two-thirds majority. For example, in the IATTC unanimity is required on any matter decided by the commission. Adoption of this voting procedure is easily understood if it is recalled that when the convention establishing the IATTC was drafted, and over the first few years that it was in force, there were only two nations involved, Costa Rica and the United States. In that case, if either nation dissented there could be no agreement of any sort. In other fisheries commissions such as the IWC and ICNAF a two-thirds majority is needed; indeed, many international organizations follow the two-thirds majority rule. In the ICCAT decisions are made on the basis of a simple majority, but there are provisions whereby a dissenting nation can postpone the date on which a provision will take effect, or even exempt themselves from compliance with the provision.

Conflicting attitudes on the issue of voting are likely to stem from the rather sharp differences in philosophies and levels of development among participants, and they could be difficult to resolve. Developing coastal RANs might logically prefer voting to take place on a simple or two-thirds majority basis because they constitute the largest voting block, and their views are often quite different from those of the technologically developed non-RANs that dominate the fisheries. The latter might prefer that decisions be unanimous so that they, in effect, would retain a veto right on any issue. To resolve these differences, more elaborate voting rules might be considered. For example, voting might be on a simple majority basis provided that all nations taking more than some specified share of the catch, say 10 percent, were included in the majority. Otherwise a substantially larger majority would be required, say three-quarters. Another possibility would be to develop some sort of weighted voting system, perhaps based on a combination of national catches and resource adjacency.

Whatever voting system is adopted, it must be understood that in most cases recommendations passed by international fishery organizations are not binding on member governments. This is true whether the voting rule is unanimity, two-thirds majority, or simple majority. Issues are dealt with in the form of resolutions which recommend action to member governments. When recommendations are approved by the assembled government representatives it is with the idea that they will be submitted to the respective governments for consideration and subsequent action. In most cases governments do accept such recommendations, but there have been instances in which the position of a nation's official representatives has differed from the position ultimately taken by their government.

Management at the Regional Level and Species Coverage

If it is best to include all interested nations, both large and small, in a global management organization, what steps might be taken to overcome the functional unwieldiness associated with having such a large voting membership? One possibility would be to divide the organization into three major regional divisions dealing with Pacific, Atlantic, and Indian Ocean fisheries, respectively. The membership of each division would consist of nations participating in the fisheries and RANs in the ocean area covered. Many nations would be represented in only one division, but some, including the most important harvesting nations, would be represented in two or even all three divisions. Many decisions, including most of those relating directly to management of the resources, could be made at the regional division level where significantly fewer nations would be involved. Other matters, those with worldwide implications, would still have to be dealt with by all members sitting as a single governing body. Such matters might include adoption of rules of procedure, the organization's fiscal affairs, interocean transfers of fleet capacity, worldwide surveillance systems, and so on. One species, the southern bluefin, would have to be treated on an interdivisional basis owing to the circumpolar distribution of the exploited stocks in the southern oceans. At the staff level, establishment of divisions for each ocean would have little direct effect —the same staff would provide advice through its directorate (or directorates) to all divisions, but it would be accountable to the parent body. Staff activities, particularly those relating to collection of fishery data, opening and closing of regional fisheries, allocation of catches, and enforcement, would be coordinated on a worldwide basis.

To further streamline organizational decision making, advisory panels could be created within each division to deal with different species groups or with major stocks of a given species exploited by essentially distinct fisheries. There is precedent for establishment of such panels in the ICNAF and in the ICCAT. The latter organization has established panels for tropical tuna (yellowfin, skipjack), northern temperate tuna (bluefin, albacore), southern temperate tuna (bluefin, albacore), and other species (bigeye, bonito, billfish, others). Membership on such panels would be voluntary. On the basis of information developed by the technical staff and from other sources, a panel would recommend management policies to the political body of the division within which it was lodged. Panels would not themselves establish policy, nor would they be technically staffed. With only directly concerned nations constituting the advisory panels, organizational decision making could be considerably expedited and much of the potential inefficiency associated with a large and diverse membership eliminated.

A global management organization such as that being described here would have responsibility for management of all species of tuna and tuna-like fish including billfish throughout the world's oceans. Each ocean division would be responsible for only those species that occurred within its own ocean. In drafting a convention creating a global organization it would be prudent to refer to general categories of fishes (such as tuna or tuna-like species) rather than to name each species of concern individually. Being too specific would reduce the flexibility of the organization and could create serious problems. For example, species not presently being harvested might become important in the future, and it would be desirable for these to be included in management's sphere of responsibility. Another possibility is that stocks of a species now considered highly migratory might upon further scientific study turn out to be quite locally distributed, falling within the juridical zones of single nations. Likewise, species now considered as resident may actually be more migratory than originally believed. However, species of interest normally would be designated in creating panels to focus on specific fisheries within the ocean divisions.

The fisheries for tuna and tuna-like species also have significant impact on other groups of animals. The most important of these are the billfish, and the various species of baitfish and porpoise. Most baitfish are taken in areas falling within territorial seas (assuming these extend out to 12 miles). Hence, harvesting of baitfish would fall under the jurisdiction of coastal states. At the request of a coastal state, however, an international organization might provide scientific and technical assistance for assessment and management of baitfish resources. Problems arising from the killing of porpoise associated with tuna in purse-seining operations in the eastern Pacific and from longline harvesting of billfish were discussed in detail in Chapter 12. It is clear that the problems of saving porpoise, harvesting tuna, and managing billfish are inextricably intertwined. Hence, it is logical that scientific studies and management analysis pertinent to these problems should be carried out jointly with overall coordination by the international organization.

Toward a Global Management System

While a global organization such as that which has been described would be the optimum vehicle for world tuna management, formation of such a body in the immediate future would be extremely difficult if not impossible considering current political and economic realities. Nations are sharply divided in their attitudes as to where benefits derived from tuna harvesting should accrue, and the extent of national jurisdiction over highly migratory pelagic species remains unresolved despite the trend toward 200-mile juridical zones.

A more reasonable expectation is that global management can evolve from the autonomous regional organizations that presently exist. These bodies are the IATTC, ICCAT, IOFC, and IPFC, and they vary greatly in their effectiveness as management bodies. None of them, however, is fully effective in all of the four basic problem areas (resource assessment, catch distribution, gear limitation, enforcement) that must be dealt with if management is to be viable over the long term. Thus the first step toward global management should be to strengthen these regional bodies. To accomplish this in the eastern Pacific and the Atlantic, present conventions need to be modified or rewritten. Perhaps the simplest approach procedurally would be to adopt protocols to the present conventions. The protocols or rewritten conventions should establish a framework for dealing effectively with the four basic problem areas and in the eastern Pacific should include provisions for porpoise protection. They would result in organizations with substantially enlarged responsibilities and powers. The requirements are similar in the Indo-Pacific area, but existing bodies in this region are limited in their capabilities and will have to be substantially reconstituted. Strengthened regional organizations would not all have to be structured in the same fashion. For example, one might have its own independent technical staff while another might depend on scientific contributions from member nations, or they might adopt different voting procedures.

Assuming that effective management bodies were functional in each ocean (or major ocean subdivision), there would be obvious reasons for them to coordinate activities such as data collection, management decision making, catch allocation, and enforcement. To provide a format within which regional organizations could mesh their management activities, a central coordinating council might be established. The exact structure of such a council would depend on the needs and desires of the regional organizations. Perhaps initially it could be composed of the chairmen of the regional organizations and their technical advisors.

While a central coordinating council would not be a policy making body in the sense of establishing or voting upon regional management programs, it could nevertheless play several important roles. For example, to prevent redundancy and inefficiency, the council could be responsible for coordinating data collection and research among the various regional organizations, regardless of whether research within the regional organizations was accomplished by independent staffs or by scientists from member nations. This could be accomplished by establishing a committee with scientific representation from the regional bodies. Another important function of the council could be to synchronize management programs so that actions taken by each regional body would be consistent with actions taken by all others. The key idea would be to prevent closed seasons, restrictions on catch, restrictions on effort, and so on, in one area from causing a transfer of vessels to areas where no restrictions were in force and where exploitation might already be intense and fleet capacity too great. Instead, transfers to areas where resources were underexploited could be encouraged. Similarly, if catches were allocated among nations in one region, options available in other regions should be considered. Requirements for enforcement would be essentially the same in all oceans and all areas, and an effective world-wide enforcement body could be established within the framework of the coordinating council. The enforcement body would have an internationally recruited staff that would not be directly controlled by any nation or regional organization, but would serve at the pleasure of all. Conflict between scientific and enforcement aspects of management could be reduced in this way.

A system under which autonomous regional organizations voluntarily coordinated their activities through a non–policy making central coordinating council would be similar in many respects to a single world tuna management organization. The main differences would be that in a single world organization:

1. The parent body would have a policy making function rather than just a coordinating function
2. Regional divisions would not be autonomous, although they would be responsible for setting policy in their respective oceans
3. A single staff would serve all regional divisions as well as the parent body.
4. Policies and procedures would be similar in the Atlantic, Pacific, and Indian Ocean divisions rather than varying from one regional organization to another.

As demands on resources become greater and greater, the need for effective management on a truly global basis will continue to increase. Under these conditions, it would be logical for cooperating regional organizations to evolve eventually into a single global organization such as has been described.

Appendix I.
Achievement of Quotas
and Fishery Closure

In discussing PAQ management incorporating yellowfin allocations, open-access fishing through international licensing, and participant fees, it was stressed that a RAN could catch none, some, or all of its allocation, or it could catch more than its allocation. The concept underlying establishment of allocations would not involve placing an upper limit on RAN yellowfin catches. While RANs would be guaranteed access to a share of the total catch, they would not be prevented from competing with other nations for a share of the catch not taken under allocations. Under these conditions, the procedure for determining the yellowfin closure date in the eastern Pacific has important implications. In closing the fishery, management must determine when the catch to date plus the catch that will be taken during the remainder of the year by RANs fishing under allocations plus incidental postclosure catches equals the overall yellowfin quota. When this point is reached, the fishery closes. The key question is whether or not RAN catches made prior to closure are considered as parts of their national allocations. In addressing this question it will be assumed that allocations are nontransferable guarantees of access rather than transferable property rights, unless otherwise stated.

Suppose that preclosure RAN catches are included in their yellowfin allocations which in turn are based on coastal zone catches by the entire international fleet (i.e., high allocations in Table 3). Non-RAN fleets would have access to the unallocated share of the catch plus any unutilized portions of RAN allocations, and in many cases their fleets would be large compared to the share of the catch available to them. On the other hand, RAN fleets in most cases would be small in comparison to the high allocations that they would receive. This means that closure would come long before any RAN (except possibly Panama) could capture its full high allocation. Of course, developing RAN fleets could continue to fish after closure until they either took their entire allocation or until the end of the year, but, in effect, their allocations would become the maximum catches available to them. The only way that a RAN could could compete with non-RANs for the unallocated catch would be to develop a fleet so large in comparison to its high allocation that it could take its whole allocation prior to closure. From that time until closure, their fleet could compete with non-RAN fleets. But development of such a large fleet would be foolish from an economic point of view because each vessel's share of the allocated catch would be reduced to such a low level that all would most likely lose money.

Still assuming that pre-closure RAN catches count against allocations, suppose that instead of high allocations, RANs receive allocations based on catches by their own fleets in their own coastal zones (i.e., low allocations in Table 3). Whether or not a RAN could take its allocation before closure would depend on the magnitude of its allocation and the size of its fleet. Nations like Panama and Nicaragua whose low allocations would be small as a result of limited coastal resources could certainly take their allocations before closure, although in some instances development of national fleets would be necessary. For nations with larger coastal resources the conclusion varies. Mexico should be able to harvest its low allocation before closure, but Ecuador probably could not since her low allocation is quite large because it is based on most of her fleet fishing the year around in her own coastal waters.

Some RANs that are unable to harvest their full allocations before closure might object to preclosure catches being treated as parts of their allocations. They could argue that allocations (even high allocations) were in effect *de facto* catch limits preventing them from competing for any share of the unallocated high-seas portion of the resource that belongs to all mankind. In addition to being entitled to an allocation related in some way to resource adjacency, in their view they should also be entitled to compete freely with all other nations for this high-seas resource. Evaluated objectively, this position has some logic, and one might conclude that RANs should not begin to fish for their yellowfin allocations until after closure. Prior to closure, RAN fleets would compete with non-RAN fleets for the portion of the catch available to all nations.

Such an arrangement in conjunction with large yellowfin allocations would be ideal from a RAN point of view, but for non-RANs the situation would be intolerable. RAN fleets would increase in size because their full guaranteed allocations would be available to them after closure. As they grew, they would provide increasingly strong competition for non-RAN fleets prior to closure. The non-RAN share of the yellowfin catch, already substantially lowered because of allocations, would be diminished even further. In fact their catches might fall to so low a level that most non-RAN vessels would be forced out of the eastern Pacific fishery altogether. It is almost inconceivable that non-RANs would agree to cooperate in such an arrangement, especially the United States, which has by far the most to lose.

Because of the economic and philosophic conflicts just outlined, some compromise on the closure issue may be essential to achievement of a viable international management program. One possible approach would be to have each RAN divide its entire fleet into two categories of vessels. One category would be designated as the "allocation fleet." All catches by vessels in this category would be counted against the nation's allocation regardless of whether they were made before or after closure. Furthermore, all vessels in a nation's allocation fleet would have to stop fishing for yellowfin when their total catch reached the allocation, even if this occurred before closure. Vessels in the second category would be designated as the "open-season fleet." Their catches would not be considered as part of a RAN's allocation, but these vessels would have to fish on strictly equal footing with all other vessels belonging to the open-season fleet, including all non-RAN vessels. That is, they would have to stop fishing for yellowfin at the time of closure even if part of the nation's allocation remained unharvested. This solution would seem to be precisely in accord with the RAN philosophy expressed earlier. Their claim to a special share of the catch based on resource adjacency is recognized, and nothing prevents them from competing with other nations for part of the international or high-seas portion of the resource. At the same time, no vessel competing for the high-seas portion of the catch would have any special advantage over other vessels such as being able to continue fishing under a RAN allocation after closure.

What might RAN strategy be under such a two-category system? Obviously, until a RAN's fleet had expanded to the point where it could harvest its entire allocation during the course of a year, all vessels would be assigned to the allocation fleet, and none would be placed in the open-season fleet. During the period of allocation fleet expansion, individual vessels should be able to operate on a very profitable basis. When their fleets had developed to the point where they could fully utilize their allocations, some RANs might choose to curtail further fleet growth, especially if allocations were at a relatively high level. If a nation did wish to develop an open-season fleet, however, it could do so, but on a strictly equal footing with all other vessels not fishing under allocations.

Whatever the method for determining which catches are to count against a RAN's allocation, if a RAN failed to harvest its full allocation in a given year, there would be no carryover into the next year. As a matter of policy, unutilized shares of allocations would be made available to the remainder of the fleet. In determining when to close the fishery, the international management agency would estimate the shares of the RAN allocations that could not be taken by the RANs themselves because of fleet limitations. Then these amounts would be added to the high-seas share of the catch that is taken prior to closure (or incidentally after closure).

So far in discussing closure procedures, RAN yellowfin allocations have been considered nontransferable guarantees of access. Suppose instead that they are treated as transferable property rights. If any RAN allocation is transferred, either in part or in whole, it should logically be treated in exactly the same way as nontransferred portions of allocations as far as closure is concerned. If a transfer is to another RAN there is no problem; the allocation of one RAN goes up and that of the other goes down. If the transfer is to a non-RAN and is substantial in comparison to the size of the non-RAN's fleet, once again there is no problem; in effect, the non-RAN is transferred into the RAN category as regards closure. However, if the allocation is transferred and made available to a non-RAN fleet that is large compared to the amount transferred, say the entire United States fleet, then an interesting situation arises. If preclosure catches count against allocations, the United States fleet would take the transferred allocation long before closure. One benefit to the United States would be extension of the open fishing period, but other non-RANs would benefit equally from this even though they did not share the transfer costs. If allocations are not taken until after closure, the large United States fleet would take the allocation rapidly after closure. Either method of closure would result in thinly spread benefits and would serve to deter the United States from seeking transfers. However, if individual boats or groups of boats could be designated as belonging to a United States "allocation fleet," then there would be a strong incentive for segments of the United States industry to bid competitively for transfer of RAN allocations.

Suppose that some nations that participate in the fishery refuse to participate in the management program. It is important to note that the estimated catches of such nations, both before and after closure, would have to be taken into account in determining the closure date in order for the total catch to equal the overall quota.

The discussion of closure has been restricted to yellowfin since this species is presently fully utilized in the eastern Pacific. If an overall quota on skipjack or some other species becomes necessary and allocations are established, then the closure method for this species should logically parallel the method adopted for yellowfin.

Appendix II.
Differential Treatment
of Yellowfin and Skipjack
under PAQ Management

In the PAQ management system described thus far there are important differences in the treatment of yellowfin and skipjack tuna. In the case of yellowfin, which is fully utilized in the eastern Pacific, there would be an overall quota and RANs would be guaranteed access to shares of the quota through a system of national allocations. The status of RANs would be further recognized by distributing a portion of participant fee receipts to them on the basis of resource adjacency. Thus, special considerations for RANs would take two forms: national allocations and adjacency-based participant fee disbursements.

In the case of skipjack, RAN claims to special consideration would be recognized initially only through adjacency-based disbursements of participant fees. There would be no adjacency-based skipjack allocations for RANs because there is no overall skipjack quota. In fact, because of open access, it is possible that effort directed at skipjack would be greater than at present, especially after the yellowfin closure and in areas where access to skipjack resources is presently limited by high license fees and national regulations. This increase in effort would be compatible with the long-range goal of increasing production from skipjack resources.

Suppose skipjack catches do increase and eventually reach a level at which scientific studies indicate a need to regulate the fishery by establishing an overall quota. The key questions are: Should RAN skipjack allocations be established? If so, upon what basis? While it may not be possible to fully resolve these questions if and when PAQ management is adopted, it is important that they be treated in general terms at the outset. If they are left open, serious future conflicts could arise that could threaten the viability of the management program.

Non-RANs might argue against skipjack allocations on the grounds that they had made significant concessions initially in agreeing to RAN yellowfin allocations with adjacency-based participant fee disbursements for both species. In return, their primary benefit is open access to yellowfin until closure and to skipjack on a year-around basis as long as no quota is necessary. In other words, open access to the entire skipjack resource is a basic, permanent condition of the original agreement, not a temporary privilege to be partially withdrawn if and when it becomes necessary to restrict the skipjack catches.

A RAN response to this argument is not hard to imagine: The only reason for not establishing skipjack allocations initially was that no need for an overall catch quota was indicated, and whenever such a need is indicated there should be no reason to treat skipjack any differently from yellowfin. On this basis, the maximum that RANs could reasonably expect to receive on the basis of their adjacency to the skipjack resource would be allocations equal to average annual catches in their coastal zones by the entire international fleet (i.e., high allocations). Minimal allocations could be based on catches made by a nation's own fleet in its own coastal zone (i.e., low allocations). These are the same limiting cases that were considered in discussing yellowfin allocation levels.

The effect of high skipjack allocations on non-RAN fleets would be even more extreme than in the case of yellowfin. For example, using data from the 1970–74 period, fully 86 percent of the skipjack catch would be allocated under this approach, a far greater figure than the 57 percent that would be allocated in the case of yellowfin (Table 3). Assuming RAN fleets developed to utilize such allocations completely, the effect would be to exclude non-RAN fleets almost totally from the skipjack fishery. Because large shares would have to be reserved for RANs after closure, the skipjack closure date would come very early in the year, long before the yellowfin closure. In fact, because postclosure incidental catches of skipjack taken with yellowfin could exceed the 14 percent of the skipjack catch that would remain unallocated, there might not be any open season for skipjack at all, even if preclosure RAN catches count against their allocations. Non-RANs could hardly agree to a management arrangement with such dire consequences as these, and RAN insistence on high skipjack allocations could dichotomize RANs and non-RANs to the detriment of rational resource management. Non-RANs might refuse to acknowledge a need for an overall skipjack quota no matter how obvious it became in order to preserve their access to the resource. Conversely, some RANs might insist on an overall quota even if the skipjack resource was clearly underexploited in order to monopolize the fishery for themselves.

In the case of low skipjack allocations, no realistic estimates of the quantities that might be allocated are possible. These figures would depend on when an overall skipjack quota became necessary and on what changes had occurred in RAN fleets and their catches in the interim. As in the case of yellowfin, however, they would be substantially lower than high allocations.

If allocations are to be established in conjunction with an overall skipjack quota, there is no reason why they could not be based on some other criteria than those used in creating yellowfin allocations. In fact, this might be preferable, considering the distributional differences between the two species. Because skipjack catches are much more concentrated in the 200-mile zone,

perhaps in establishing allocations there should be more emphasis on historical catches and harvesting capability and less on the magnitude of adjacent
resources. Another important difference is that yellowfin in the eastern Pacific
appear to be largely confined to that ocean area while skipjack apparently
migrate to the eastern Pacific from the central Pacific where they occur within
200 miles of various nations. The rights and prerogatives of these central
Pacific RANs should be considered in allocating the eastern Pacific skipjack
resource. If both yellowfin and skipjack are subject to overall quotas with
partial catch allocations, estimating postclosure catches of the two species so
that closure dates can be determined will be a complex problem that the
international management body would have to resolve.

In searching for solutions to the difficult problems involved in management
and utilization of tuna resources, it is impossible to distribute the burden
equally at all times, although hopefully the bulk of it will not always fall on
the same nation or group of nations. In the case of an open-access skipjack
fishery in which there would be no overall quota and no RAN allocations, the
burden would probably fall most heavily on Ecuador. Skipjack catches on the
average are larger off her coasts than off any other nation (Table 2). Also
Ecuador has developed a fair-sized fleet of small bait boats and seiners. Most
of these vessels are restricted to fishing inshore waters near their home ports,
primarily for skipjack. The Ecuadorian government has provided this fleet
with protection from foreign competition by charging high license fees to
foreign vessels wishing to fish in Ecuadorian waters, and in the past relatively
few boats have purchased licenses. Also, under present regulations, those that
purchase licenses must fish at least 60 miles from shore. With open access,
the internaational fleet fishing for skipjack in Ecuadorian waters would increase in size and would provide competition for small Ecuadorian boats.
Thus, Ecuadorian gains resulting from its yellowfin allocation and from
participant fee distributions for both species might be offset by losses owing
to increased competition for skipjack. No other RAN is in quite this position.

Assuming that other nations involved in the eastern Pacific tuna fishery
were in agreement on open-access management with partial allocation of an
overall quota and participant fees, what choices would Ecuador have? Basically there would be two. Ecuador could decide that the advantages of participating outweighed the disadvantage of increased competition for skipjack.
Or she could decide to go it alone and exercise unilateral control over all
fishing activities in her 200-mile zone. Under the circumstances, it would not
be surprising if Ecuador chose the latter alternative and refused, at least
initially, to participate in an international management agreement. But at
least two factors could mitigate against this choice. First, Ecuador's fleet is
steadily changing from one largely restricted to local grounds to one made
up of farther-ranging vessels which, without an international license, could
not fish in the coastal zones of other RANs. Second, territorial waters within

12 miles would remain closed to foreign flag fishing under international PAQ management, and Ecuador is relatively unique in that when skipjack are abundant off her shores, rather substantial catches are taken within her 12-mile zone (Table 4). Thus, despite drawbacks, the most reasonable available option for Ecuador might be to cooperate in a PAQ management agreement. Whatever her decision, it is important that any international PAQ management agreement be left open to adherence by nations that are not original participants, just as in the case of the convention creating the IATTC.

Problems associated with differential management of yellowfin and skipjack have been discussed in detail because these are the two most important species in the eastern Pacific fishery. However, the discussion also illustrates the point that no single management formula can apply universally to all species, fisheries, or user groups. In every situation, specific arrangements will depend on the unique circumstances involved. Management of less important species in the eastern Pacific such as bigeye and northern bluefin and management of longline fisheries are further examples of challenging, multispecies management problems. Closely related are management problems related to multiple user groups. Examples include the competition between purse seiners and longliners that harvest yellowfin at different ages in the Atlantic, and disputes between sport-fishing and commercial fishing interests over the harvest of billfish (see Chap. 12) and bluefin tuna. In each situation, all legitimate interests must be recognized, alternative courses of action considered, and some solution (probably a compromise) agreed upon if the resources involved are to be successfully managed.

Appendix III.
Changes in Fleet Composition:
Development of RAN Fleets

Any international management system involving establishment of substantial adjacency-related catch allocations for RANs will stimulate significant changes in the composition of the international fleet. If RANs are allocated shares of the catch, it is likely that they will want to expand their fleets until they are large enough to fully harvest their allocations on a reasonably efficient basis. Vessels can enter developing RAN fleets in various ways: Vessels presently participating in the fishery can change flags, or vessels can enter the fishery for the first time either by being newly built or by transferring from other fisheries. These alternatives will be explored in this appendix.

Other changes in fleet composition include entrance of vessels into fleets of non-RANs already participating in a fishery, entrance of fleets from non-RANs not presently active in the fishery, and entrance of vessels into RAN fleets that are fully developed in the sense that they are large enough to harvest their entire annual allocations. If a fishery is underdeveloped in the sense that the resource is not being fully utilized, then from the point of view of current fleet economics there is less compelling reason to restrict entry of vessels. But if the international fleet participating in a fishery is already large enough to fully utilize the available resources, then entry of vessels in the above categories should be carefully controlled and further fleet expansion discouraged. This is the case in the eastern Pacific and other important tuna fisheries. The problem of controlling fleet size is covered in Appendix IV.

Pros and Cons of Flag Changes versus New Construction

In considering entry into underdeveloped RAN fleets, several types of flag change must be distinguished. Intraregional flag changes from one underdeveloped RAN fleet to another need merely to be noted. These would have no net effect on the functioning of the overall system and would normally be uncommon. Interregional flag changes could be significant in some cases (e.g., the possible transfer of United States flag vessels to fisheries in other ocean regions) and will be touched on later. Of much greater potential importance are intraregional flag changes from non-RAN fleets or fully developed RAN fleets to underdeveloped RAN fleets. These will be emphasized in the remainder of this section, and the term "flag changes" will refer to transfers of this type, unless stated otherwise. The discussion will focus on the present situation in the eastern Pacific, where such transfers could be very important, especially those involving United States vessels. Also, because fleet transition problems would be most serious in the case of high allocations, this situation is assumed throughout much of this appendix. (This should not be construed as an endorsement of high allocations.) If allocations are much lower, RAN fleet growth would be more limited and associated problems minimized.

The decision to transfer a vessel into a developing RAN fleet would be voluntary, with the prime objective for changing flags being to fish under a RAN allocation that is large compared to the size of the RAN's fleet. Under such conditions, a flag change could mean reduced competition for the vessel and improved access to resources. Certain factors, however, could mitigate against flag changes. For example, a RAN might impose special requirements such as that part of the crew must hold RAN citizenship, or that there must be some RAN participation in vessel ownership, or that some part of the catch must be landed at local processing facilities. To an individual owner, such requirements could range from being highly onerous to being fully acceptable. Other factors that could discourage flag changes might be political instability in a RAN, risk of future nationalization, excessive taxation, and so on. Finally, an owner might have strong nonbusiness reasons for preferring his present flag (citizenship, place of residence, family tradition, etc.). The important point is that in each instance it is the individual owner who must weigh the pros and the cons. If he decides to change flags, the risks are his to bear and the potential benefits his to reap.

Flag changes have one important advantage over RAN fleet development by construction of new vessels, at least in the eastern Pacific. They do not increase total fleet size whereas new construction does. Since the eastern Pacific fleet is already larger than necessary, this is of significance to all non-RANs, but especially to the United States, because most flag changes would involve transfers from her fleet. Provided that preclosure RAN catches count against RAN allocations, reduction of the non-RAN fleet competing for the unallocated share of the catch means a longer open fishing period and greater catches prior to closure for remaining non-RAN vessels. The economic benefits to the non-RAN fleet are obvious and will be quantitatively illustrated shortly by means of an example.

Until their fleets became fully developed, flag changes would not offer RANs any advantage in terms of catch per vessel over new construction. Flag changes could result, however, in more timely fleet growth because RANs could avoid the potentially severe problem of finding investment capital, and construction time lags could be eliminated. Also if a RAN ever did want to fish competitively for a share of the unallocated catch, it would benefit from the total fleet being smaller in the same way that non-RANs would benefit. For those reasons RANs might generally favor fleet development via the flag changes over new construction. It is unlikely, however, that they would support any agreement totally disallowing new construction as a means for RAN fleet growth. If the alternative of new construction were to be eliminated, then at least some of the incentive for non-RAN vessels to make flag changes would be lost. Hence, RANs would want to retain the new construction option. Also, RANs wishing to support or develop shipbuilding industries might favor new construction.

Interregional transfers of vessels into underdeveloped RAN fleets from other fisheries would be equivalent to construction of new capacity from the point of view of the region into which they transferred, at least in the sense that such transfers would increase regional fleet capacity. Thus, if RAN fleet growth by new construction is to be discouraged in order to minimize total fleet growth, then transfers from other regions into underdeveloped non-RAN fleets should be similarly discouraged, except perhaps in cases not involving flag changes (e.g., in the case of France in the eastern Pacific).

In spite of arguments favoring RAN fleet development by flag changes, there are several reasons why flag changes might be opposed. No nation which has developed a large and technologically sophisticated fishing industry is likely to be pleased with the prospect of a large segment of its fleet transferring to foreign flags (with concomitant reduction in its share of the total catch), even if there are sound political and economic reasons for such changes. For this reason, internal political opposition can be expected, and this factor could take on even greater significance at the regional and local levels (e.g., in southern California, especially San Diego, or Puerto Rico, especially Mayaguez and Ponce).

Additional opposition to flag changes, at least in the United States, could be expected from the shipbuilding industry (both management and unions) and from tuna fishermen themselves acting through their unions. Shipbuilder opposition would be understandable, but the question could be raised as to whether their short-term loss of business might not be more than balanced in the long run by an improved economic climate in the fishing industry where a smaller international fleet spells better fishing. The concern of fishermen would be based on the loss of a large number of jobs to non–United States labor. And fishing being what it is, the loss would involve more than just a job for many; it would involve a way of life. One could certainly sympathize with the fishermen, but at least for those on boats still under the United States flag, fishing should be better than it would otherwise be.

A more subtle form of opposition to flag changes could arise within the community of United States boat owners. Within this group a wide range of attitudes could exist. Some might be anxious to change flags in hopes of gaining a competitive advantage even if there were adverse aspects and considerable risks involved. Others would be reluctant, but they would do so for sufficient justification. Some, perhaps those with the deepest roots in the California-based fishery, might prefer to withdraw from the business rather than change flags. So far it has been assumed that flag changes would be voluntary and unimpeded by the nation involved. But human nature being as it is, things might not work out quite so neatly. Boat owners not electing to make flag changes themselves might well be vigorously opposed to allowing competing boat owners the same opportunity for fear that these competitors might derive a significant advantage. Such conflicts within the intensely competitive United States industry could complicate and perhaps even deadlock negotiations leading to an international management agreement.

It is of interest to explore the consequences of restricting flag changes, perhaps through stipulations incorporated into an international agreement establishing a PAQ management system. Relying only on new construction, how rapidly might RAN fleets grow? It is difficult to answer this question, but in view of the economic problems confronting the already large international fleet, it is tempting to speculate that investment capital might be rather scarce with the result that further RAN fleet growth would be slow. This thinking implies that large shares of the RAN allocations could not be taken by their existing fleets and would remain available to the non-RAN fleets. In other words, the non-RAN share of the total catch would be reduced very slowly, and the excess capacity problem would not be overly aggravated.

If the above reasoning is sound, then for non-RANs it would constitute a strong argument for restricting flag changes. (RANs, of course, would see it as an argument for allowing flag changes.) But if the reasoning is unsound, and RAN fleets grew fairly rapidly by means of new construction, then restricting flag changes would be a questionable policy, especially for non-RANs whose share of the catch would decline rapidly with no concomitant reduction of their fleets. While the availability of capital for construction of RAN fleets is clearly the critical issue, and there is one good reason to believe that investors might make capital more available than in the past. In recent years factors such as uncertainty associated with extension of jurisdiction to 200 miles, the likelihood of extreme future competition for the resource, and the absence of adjacency-related national allocations have probably combined to strongly inhibit investment in new RAN construction. But establishment of allocations and resolution of the jurisdictional problems under an international agreement would alleviate these inhibiting factors. It seems likely that this could reverse investor thinking and draw money into construction of new and modern RAN vessels. Hence the argument that restriction of flag changes would greatly inhibit RAN fleet growth may well be unsound.

TABLE 13.3 POSSIBLE METHODS FOR PHASING IN NATIONAL YELLOWFIN ALLOCATIONS IN THE EASTERN PACIFIC OVER A 10-YEAR PERIOD

FAST PHASE-IN (all figures in thousands of short tons)

Nation	Year 0 catch	Annual increment	Annual allocations during ten-year phase-in period									
			1	2	3	4	5	6	7	8	9	10
Colombia	-	0.61	0.6	1.2	1.8	2.4	3.0	3.7	4.3	4.9	5.5	6.1*
Costa Rica	0.5	2.27	2.8	5.0	7.3	9.6	11.9	14.1	16.4	18.7	20.9	22.7*
Ecuador	8.6	1.63	10.2	11.8	13.4	15.1	16.3*	16.3	16.3	16.3	16.3	16.3
El Salvador	-	0.36	0.4	0.7	1.1	1.4	1.8	2.2	2.5	2.9	3.2	3.6*
France (Clipperton)	2.0	0.38	2.4	2.7	3.1	3.5	3.8*	3.8	3.8	3.8	3.8	3.8
Guatemala	-	0.41	0.4	0.8	1.2	1.6	2.0	2.5	2.9	3.3	3.7	4.1*
Mexico	14.5	3.60	18.1	21.7	25.3	28.9	32.5	36.0*	36.0	36.0	36.0	36.0
Nicaragua	-	0.17	0.2	0.3	0.5	0.7	0.8	1.0	1.2	1.3	1.5	1.7*
Panama	6.9	0.80	7.7	8.0*	8.0	8.0	8.0	8.0	8.0	8.0	8.0	8.0
Peru	1.5	0.66	2.2	2.8	3.5	4.2	4.8	5.5	6.1	6.6*	6.6	6.6
U.S.A.	141.6	-	0.009*	0.009	0.009	0.009	0.009	0.009	0.009	0.009	0.009	0.009
RAN total (U.S.A. excluded)	34.0											
Total including U.S	175.6	Allocated	44.9	55.3	65.3	75.4	85.0	93.0	97.5	101.7	105.5	108.8
Bermuda, Canada, Japan, Netherlands Antilles, Spain	16.8	Unallocated	147.5	137.1	127.0	116.9	107.4	99.4	94.9	90.7	86.8	83.5
Grand total	192.4		192.4	192.4	192.4	192.4	192.4	192.4	192.4	192.4	192.4	192.4

INTERMEDIATE PHASE-IN (all figures in thousands of short tons)

Nation	Year 0 catch	Annual increment	Annual allocations during ten-year phase-in period									
			1	2	3	4	5	6	7	8	9	10
Colombia	-	0.61	0.6	1.2	1.8	2.4	3.0	3.7	4.3	4.9	5.5	6.1*
Costa Rica	0.5	2.22	2.7	4.9	7.2	9.4	11.6	13.8	16.0	18.3	20.5	22.7*
Ecuador	8.6	0.77	9.3	10.1	10.9	11.7	12.4	13.2	14.0	14.7	15.5	16.3*
El Salvador	-	0.36	0.4	0.7	1.1	1.4	1.8	2.2	2.5	2.9	3.2	3.6*
France (Clipperton)	2.0	0.18	2.2	2.3	2.5	2.7	2.9	3.1	3.2	3.4	3.6	3.8*
Guatemala	-	0.41	0.4	0.8	1.2	1.6	2.0	2.5	2.9	3.3	3.7	4.1*
Mexico	14.5	2.15	16.6	18.8	20.9	23.1	25.2	27.4	29.6	31.7	33.9	36.0*
Nicaragua	-	0.17	0.2	0.3	0.5	0.7	0.8	1.0	1.2	1.3	1.5	1.7*

	Year 0 catch	Annual increment	1	2	3	4	5	6	7	8	9	10
Panama	6.9	0.11	7.0	7.1	7.2	7.4	7.5	7.6	7.7	7.8	7.9	8.0*
Peru	1.5	0.50	2.0	2.5	3.0	3.5	4.0	4.5	5.0	5.6	6.1	6.6*
U.S.A.	141.6	-	0.009*	0.009	0.009	0.009	0.009	0.009	0.009	0.009	0.009	0.009
RAN total (U.S.A. excluded)	34.0											
Total including U.S	175.6	Allocated	41.5	48.9	56.4	63.9	71.4	78.9	86.4	93.9	101.3	108.8
Bermuda, Canada, Japan, Netherlands Antilles, Spain	16.8	Unallocated	150.9	143.4	135.9	128.4	120.9	113.5	106.0	98.5	91.0	83.5
Grand total	192.4		192.4	192.4	192.4	192.4	192.4	192.4	192.4	192.4	192.4	192.4

SLOW PHASE-IN (all figures in thousands of short tons)

Nation	Year 0 catch	Annual increment	Annual allocations during ten-year phase-in period									
			1	2	3	4	5	6	7	8	9	10
Colombia	-	0.61	0.6	1.2	1.8	2.4	3.0	3.7	4.3	4.9	5.5	6.1*
Costa Rica	0.5	2.27	2.3	4.5	6.8	9.1	11.4	13.6	15.9	18.2	20.4	22.7*
Ecuador	8.6	1.63	8.6	8.6	8.6	8.6	8.6	9.8	11.4	13.0	14.7	16.3*
El Salvador	-	0.36	0.4	0.7	1.1	1.4	1.8	2.2	2.5	2.9	3.2	3.6*
France (Clipperton)	2.0	0.38	2.0	2.0	2.0	2.0	2.0	2.3	2.6	3.0	3.4	3.8*
Guatemala	-	0.41	0.4	0.8	1.2	1.6	2.0	2.5	2.9	3.3	3.7	4.1*
Mexico	14.5	3.60	14.5	14.5	14.5	14.5	18.0	21.6	25.2	28.8	32.4	36.0*
Nicaragua	-	0.17	0.2	0.3	0.5	0.7	0.8	1.0	1.2	1.3	1.5	1.7*
Panama	6.9	0.80	6.9	6.9	6.9	6.9	6.9	6.9	6.9	6.9	7.2	8.0*
Peru	1.5	0.66	1.5	1.5	2.0	2.6	3.3	3.9	4.6	5.2	5.9	6.6*
U.S.A.	141.6	-	0.009*	0.009	0.009	0.009	0.009	0.009	0.009	0.009	0.009	0.009
RAN total (U.S.A. excluded)	34.0											
Total including U.S	175.6	Allocated	37.3	41.1	45.4	49.8	57.8	67.4	77.5	87.6	98.0	108.8
Bermuda, Canada, Japan, Netherlands Antilles, Spain	16.8	Unallocated	155.1	151.2	147.0	142.5	134.5	125.0	114.9	104.8	94.4	83.5
Grand total	192.4		192.4	192.4	192.4	192.4	192.4	192.4	192.4	192.4	192.4	192.4

NOTE: Final allocations are indicated by asterisks in the year in which they are reached.
Sums of individual allocations may not agree exactly with totals shown because of rounding.

Effect of Transition to PAQ Management on RAN and Non-RAN Fleets

Regardless of whether RAN fleets grow as a result of new construction, through flag changes, or by some combination of these two methods, the rapidity of transition to PAQ management would have important economic consequences for both RANs and non-RANs. To evaluate these possible consequences, high allocations based on catches by the entire international fleet in national 200-mile zones were assumed because transition problems would be most extreme in this case. Several alternative plans for phasing in RAN allocations ranging from as rapidly as possible to rather gradually were examined to determine their effect on the patterns of change in fleet composition and catch distribution. For each allocation phase-in alternative two extremes for RAN fleet growth were evaluated: all growth by flag changes and all growth by new construction. For RAN fleet growth by a combination of these two methods, results would be intermediate.

The most rapid transition would take place if national yellowfin allocations became fully available to RAN fleets immediately upon conclusion of an agreement. This will be referred to as the "immediate phase-in" case. Alternatively, allocations could be gradually phased in over a period of years, and three gradual phase-in schemes were defined for detailed consideration. In doing so, a ten-year phase-in period was assumed, and a basic requirement was that a nation's initial year-1 yellowfin allocation should not be less than its preceding year-0 catch (unless its year-0 catch already exceeded its final allocation). The three schemes are referred to as fast, intermediate, and slow phase-in, and can best be understood by examining Table 13, which shows changes in allocations over the ten-year phase-in period for individual nations. This table and the rest of the tables and figures in this section can be thought of as a continuation of the PAQ management examples presented in Table 5. The specific rules for phase-in allocations are as follows:

1. Fast phase-in: The annual increment in each nation's allocation is one-tenth of its final allocation. This amount is added to the previous year's allocation (or in year 1, to the year-0 catch) until the final allocation is reached. Note that allocations of all nations begin increasing right away and that the year in which a nation reaches its final allocation depends on its year-0 catch (compare Costa Rica, Mexico, and Panama).

2. Intermediate phase-in: The annual increment is one-tenth of the difference between the final allocation and the year-0 catch. It is added to the previous year's allocation (or to the year-0 catch) until the final allocation is reached in year 10.

3. Slow phase-in: The annual increment is one-tenth of the final allocation. These amounts accumulate until they exceed a nation's year-0 catch, after which they determine the nation's allocation. Prior to this, the allocation equals the year-0 catch. Note that allocations for nations without fisheries increase right away while allocations of nations with partially developed fisheries do not increase for several years (compare Colombia and Ecuador).

In the three phase-in schemes of Table 13, year-0 catches correspond to yellowfin catches in recent years as given in the first example of Table 5, while the final high yellowfin allocations, whether they are reached in year 10 or earlier, correspond to catches given in the second example of Table 5 where it is assumed that all nations fully utilize their high allocations. The terms *fast, intermediate,* and *slow* refer to the rates at which the total amounts allocated increase during the early years of transition. By year 10, individual national allocations reach the same final levels under all phase-in schemes.

Given high allocations, a rationale could be developed to justify any of the three gradual phase-in alternatives. For example, it seems reasonable that if the maximum phase-in period is to be 10 years, the annual increment should be 10 percent of the final allocation. But if incrementation begins from year-0 catch levels (fast phase-in), a RAN with a partially developed fleet would reach its full allocation before RANs without fleets. To some this might seem inequitable. After all, why should any RAN have to wait longer than others to derive full benefits from its allocation? If all RANs are to reach their final allocation in year 10 using 10 percent increments, then any increase in a RAN's allocation must wait until the accumulating 10 percent increments exceed the nation's year-0 catch (slow phase-in). But this approach would interrupt the progress of RANs with partially developed fleets, which might seen unjustified. Considering both points of view, a compromise allowing for steady but more moderate increments for RANs with partially developed fleets might be a more readily acceptable solution (intermediate phase-in).

Changes in combined RAN yellowfin allocations for all phase-in alternatives are plotted in Figure 13. Differences among the four possibilities during the transition period are clearly significant even though they all lead to a common result in year 10. For example, at the 5-year halfway point, allocations could be anywhere from 58,000 to 109,000 short tons (i.e., from 53 percent to 100 percent, respectively, of the total to be allocated, or from 30 percent to 57 percent of the entire eastern Pacific yellowfin resource). Because the slow or intermediate transition alternatives initially result in less abrupt change than fast or immediate transition, they should probably be favored. Also, it should be noted that differences among the three gradual phase-in alternatives primarily involve nations with partially developed fleets (Table 13). Allocation phase-in patterns would be the same for nations without fleets under all three

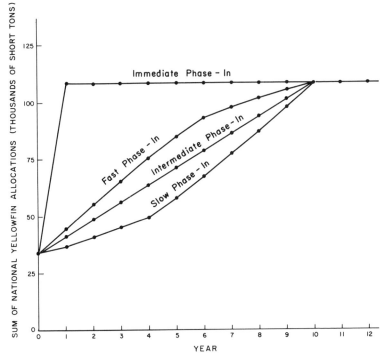

Figure 13. Four alternatives for phasing in combined national yellowfin allocations in the eastern Pacific. (Phase-in methods are fully described in the text)

alternatives, and a nation with a fully developed fleet would also be unaffected by the choice.

Before examining changes in fleet composition that might result from implementing a system of high RAN yellowfin allocations, some assumption must be made on how skipjack catches would be divided. It is uncertain whether or not an overall quota on skipjack will ever be necessary. Also, if a skipjack quota is imposed, its size and the extent of skipjack allocations are unknown. Therefore, two possibilities are considered which correspond to probable minimization and maximization of RAN skipjack catches (and vice-versa for non-RANs). In the first case it is assumed that there would be no skipjack allocations, and that RAN fleet growth would be determined by RAN yellowfin allocations. It is further assumed that each RAN's skipjack catch would be proportional to the size of its fleet as compared to the entire international fleet. It seems unlikely that RAN fleets would take a smaller share of the total skipjack catch than that which would be taken under these assumptions. For example, in 1974 RAN fleets constituted about 17 percent of total carrying capacity, yet they took about 24 percent of the total eastern Pacific skipjack catch. Hence, under the first case RAN skipjack catches can be considered minimized. In the second case it is assumed that national skipjack allocations based on coastal zone catches of the entire international fleet would be established and phased in just as in the case of yellowfin. On this basis 86 percent of the skipjack catch would be allocated as can be seen in Table 3. It is then assumed that RAN fleet growth would be determined by combined yellowfin and skipjack allocations. In this case RAN skipjack catches can be considered maximized. In both cases it is assumed that the total skipjack catch equals the average skipjack catch during the 1970–74 period.

To summarize, there are two types of fleet growth (by flag changes or by new construction), four allocation phase-in alternatives (immediate, fast, intermediate, and slow), and two ways of treating skipjack (no RAN allocations or high RAN allocations) to be considered. This gives sixteen combinations, as indicated in Table 14.

As allocations are phased in, RAN fleets would grow and their catches would increase. For all sixteen cases the size of the combined RAN fleet during each year, its annual catch of both species combined, and its annual catch per capacity ton were calculated. Similar calculations were made for the combined non-RAN fleet. In all cases, increases in RAN carrying capacity were assumed to be sufficient to take added catch increments at the rate of 3 tons annually per ton of added carrying capacity (e.g., a 1,000-ton seiner added to the fleet would take 3,000 tons per year). At this rate of production a modern purse seiner should be able to operate on a reasonably profitable basis. Depending on the case in question, various assumptions were made concerning time lags involved in making flag changes and building new vessels (see Table 14). In all cases RAN and non-RAN fleets in year 0 were assumed to be equivalent to their size during the last quarter of 1974.

Results of these calculations are summarized in Figure 14 and Table 15. Figure 14 shows transitional changes in RAN and non-RAN fleet capacities, annual catches, and catches per capacity ton in cases involving yellowfin allocation only (cases 1–8 of Table 14). Table 15 indicates the situation at the end of the ten-year phase-in period for all 16 cases. Note that there are no differences after ten years owing to phase-in alternative among otherwise similar cases (i.e., cases 1–4, 5–8, 9–12, and 13–16).

For flag changes, Figure 14 shows increases in RAN fleets corresponding to declines in non-RAN fleets. If RAN fleet growth is by new construction, non-RAN fleets remain unchanged as RAN fleets develop during years 3 through 10. If both species are allocated, RAN fleets and catches after 10 years are identical for either method of fleet development (Table 15). But if only yel-

TABLE 14

DEFINITION OF 16 HYPOTHETICAL CASES EXAMINING CHANGES IN RAN AND NON-RAN FLEET CARRYING CAPACITIES, TOTAL CATCHES, AND CATCHES PER CAPACITY TON OVER A 10-YEAR ALLOCATION PHASE-IN PERIOD

Case number	High adjacency based allocations of—	RAN fleet growth by—	Allocation phase-in rate	Time lag assumption
1.	Yellowfin only	Flag changes	Immediate	A
2.			Fast	B
3.			Intermediate	B
4.			Slow	B
5.		New construction	Immediate	C
6.			Fast	E
7.			Intermediate	D
8.			Slow	D
9.	Yellowfin and skipjack	Flag changes	Immediate	A
10.			Fast	B
11.			Intermediate	B
12.			Slow	B
13.		New construction	Immediate	C
14.			Fast	E
15.			Intermediate	D
16.			Slow	D

Assumptions concerning time lags involved in making flag changes or for new construction:

A. Transfer by flag changes of capacity sufficient to enable RANs to harvest the entire RAN allocation requires 2 years. Half of the transfers from non-RAN to RAN fleets takes place in year 1 and half in year 2.

B. In each year capacity sufficient to harvest the RAN allocation increase is transferred from non-RAN fleets to RAN fleets by means of flag changes.

C. The minimum time lag for financing and constructing new vessels is 3 years. Hence there are no changes in RAN fleets in years 1 and 2. Capacity sufficient to enable RANs to harvest the entire RAN allocation is added through new construction in years 3–5, with one third of the new capacity being added in each year.

D. As under assumption C, no new construction is completed before year 3. In each year from year 3 on, new capacity sufficient to harvest the RAN allocation increase (or accumulation of increases) is added to RAN fleets through new construction.

E. Same as assumption D except that in any given year RAN capacity is not allowed to exceed the level of capacity reached in the corresponding immediate phase-in case (i.e., case 5 or 13). (This affects only year 3 in cases 6 and 14.)

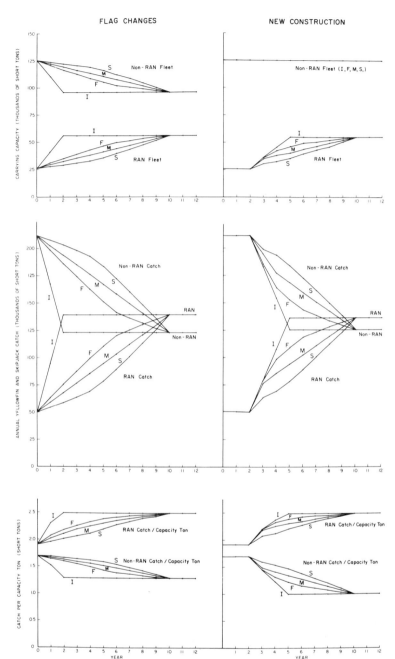

Figure 14. Changes in RAN and non-RAN fleet carrying capacities, catches, and catches per capacity ton over a 10-year allocation phase-in period. Cases 1–8 of Table 14 are covered in which there are high allocations of yellowfin only. I, F, M, and S stand, respectively, for immediate, fast, intermediate, and slow allocation phase-in.

lowfin are allocated, flag changes ultimately result in slightly larger RAN fleets than those that would be achieved with new construction. This is the case because RAN fleets developed through flag changes would constitute a larger proportion of the total fleet (whose size is fixed) and therefore take a somewhat larger share of the skipjack catch. Interim year differences in fleets associated with different phase-in alternatives are substantial, especially in comparisons involving immediate phase-in. For example, in the case of RAN fleets developing through flag changes, the difference between slow and immediate phase-in amounts to nearly half of the final fleet size in some years.

As RAN fleets grow, the RAN share of the catch also grows, while the non-RAN share declines. By the end of the phase-in period RAN catches exceed non-RAN catches regardless of whether RAN fleet growth results from flag changes or new construction. Interim year catches vary greatly depending on the method of allocation phase-in, especially if fleet growth is by flag changes. In this case the difference between slow and immediate phase-in exceeds 50 percent of the ultimate RAN catch in certain years.

The lower plots in Figure 14 show changes in the annual catch per ton of carrying capacity for cases involving yellowfin allocations only. For the RAN fleet these figures increase as allocations are phased in because RAN catches increase relatively more rapidly than RAN fleet size. This results from the assumption that vessels entering RAN fleets harvest tuna at an annual rate of 3 tons per ton of capacity, which is greater than the rate for the existing RAN fleet. The method of RAN fleet expansion has no significant effect on the final catch per capacity ton because fleet growth by either flag changes or new construction results in nearly equal catches and fleet capacities. Non-RAN catch per capacity ton drops in both cases, but it drops to a substantially lower level in the new construction case (1.29 tons for flag changes versus 1.00 tons for new construction). This difference is due to the fact that with flag changes non-RAN fleets decline as their catches decline, while with new construction their declining catches are taken by a fleet of fixed size. Catch per capacity ton for the combined RAN and non-RAN fleet will also obviously be substantially higher if RAN fleet growth results from flag changes rather than from new construction. As noted earlier, these results provide a significant reason for favoring RAN fleet growth by flag changes.

In cases 9–16 of Table 14, high allocations of both yellowfin and skipjack resources are assumed. These cases are of interest because they define the most extreme changes that can reasonably be postulated. The general patterns of change would be similar to those plotted in Figure 14 for cases involving only yellowfin allocation, but all changes would be magnified, as indicated in Table 15. RAN fleets would grow even larger, and, in the case of flag changes, non-RAN fleets would be even more diminished. Similarly, RAN catches would increase further and non-RAN catches decline more drastically. The annual catch per capacity ton would be somewhat higher for RANs if both species

TABLE 15

FINAL RAN AND NON-RAN FLEET CARRYING CAPACITIES, TOTAL CATCHES, AND CATCHES PER CAPACITY TON FOR ALL 16 ALLOCATION PHASE-IN CASES OF TABLE 14

Fleet Capacity (thousands of short tons)

Fleet growth by—		1974 (last quarter)	After ten-year allocation phase-in period	
			Yellowfin allocations only	Yellowfin and skipjack allocations
Flag changes	RANS	26.4	55.9	65.8
	non-RANS	125.0	95.5	85.6
	Total	151.4	151.4	151.4
New construction	RANS	26.4	55.0	65.8
	non-RANS	125.0	125.0	125.0
	Total	151.4	180.1	190.8

Annual Yellowfin and Skipjack Catch (thousands of short tons)

Fleet growth by—		Average, 1970–74	After ten-year allocation phase-in period	
			Yellowfin allocations only	Yellowfin and skipjack allocations
Flag changes	RANs	50.5	138.9	168.6
	non-RANs	211.6	123.2	93.5
	Total	262.1	262.1	262.1
New construction	RANs	50.5	136.5	168.6
	non-RANs	211.6	125.6	93.5
	Total	262.1	262.1	262.1

Catch per Capacity Ton (short tons)

Fleet growth by—		1970–74	After ten-year allocation phase-in period	
			Yellowfin allocations only	Yellowfin and skipjack allocations
Flag changes	RANs	1.91	2.49	2.56
	non-RANs	1.69	1.29	1.09
	Entire fleet	1.73	1.73	1.73
New construction	RANs	1.91	2.48	2.56
	non-RANs	1.69	1.00	.75
	Entire fleet	1.73	1.46	1.37

were allocated, but for non-RANs the figures would be sharply lower, in the case of flag changes 1.09 tons rather than 1.29 tons, and in the case of new construction 0.75 tons rather than 1.00 tons.

While this example brings out the significant differences between allocation phase-in methods and illustrates the advantage in terms of catch per capacity ton of RAN fleet growth by flag changes rather than by new construction, perhaps its most important function is to highlight the problems facing non-RAN fleets if high adjacency-based allocations are established—even if only for yellowfin. The non-RAN catch per capacity ton given in the recent years column of Table 15 is 1.69 tons. At this level of production in the eastern Pacific, some vessels have already experienced difficulties, especially newer vessels financed under loan agreements that now require large principal and interest payments. Under the example's assumptions, a sharp decline in this already depressed non-RAN catch per capacity ton figure is indicated, with the minimum decline being 24 percent to 1.29 tons in the case of flag changes with yellowfin allocations only. A reduction in productivity to this level would be catastrophic economically for non-RAN vessels if it were to actually occur, but fortunately there are some mitigating factors.

Factors Mitigating the Impact of PAQ Management

In negotiations to establish a PAQ management system much could be done to alleviate the serious problem of declining non-RAN vessel productivity. First, yellowfin allocations could be set at levels below those assumed in the fleet transition example, which are the maximum that could be rationally advocated. Likewise, if skipjack allocations ever become necessary, they could be set below the high levels assumed in the example. Also, RAN fleet growth by flag changes could be encouraged, rather than growth by new construction. Another possibility would be to select a slow or intermediate allocation phase-in plan rather than a fast or immediate phase-in plan. This would minimize the initial impact of allocation.

All of the above measures are suggested by the fleet transition example itself, but because it was necessarily based on a number of simplifying assumptions, there are several additional factors, not accounted for in the example, that would tend to rather significantly reduce the impact of allocation on non-RAN fleets. For one thing, when vessels presently fishing in the eastern Pacific leave the non-RAN fleet through sinking or obsolescence or by transferring to fisheries in other regions, some may not be replaced. This would make it possible for productivity of remaining vessels to increase. Additionally, in certain years part of the non-RAN fleet fishes outside the eastern Pacific area after closure (mainly in the eastern Atlantic). Catches made in these areas are not included in either the catch or catch per capacity ton calculations of the example. Also excluded from catch and annual catch per capacity ton calculations are catches of species other than yellowfin and skipjack. These species include bigeye, northern bluefin, bonito, albacore, and black skipjack, which combined produce annual catches in the 20,000–30,000 ton range. Another possibility is that non-RAN vessels could increase their skipjack catches to compensate for yellowfin losses caused by the establishment of allocations. This could result partly from gaining open access to all areas as a concomitant of establishing allocations and partly from improved skipjack fishing techniques.

Another important mitigating factor is that certain RANs might not fully utilize their allocations. Any unutilized shares of allocations would remain available to non-RAN fleets (assuming allocations are nontransferable guarantees), and would reduce non-RAN per vessel productivity losses. If significant quantities of fish were involved, this factor alone could greatly reduce the adverse impact of establishing an allocation system. This suggests the interesting possibility of providing special incentives for RANs to refrain from harvesting their allocations. One approach might be to increase participant fee disbursements to RANs that agree to refrain from harvesting all or part of their allocations. Funds that would otherwise go to harvesting nations on the basis of their catches in the final stage of the participant fee redistribution system described in Chapter 8 could be used for this purpose. Any such incentive payments to RANs should probably only be made for relatively long-term commitments to refrain from harvesting, say, ten years. In other words, a RAN should not receive increased disbursements over the short term simply because its present fleet is inadequate to harvest its full allocation. If allocations were treated as transferable property rights, then a RAN would also have to agree not to transfer its allocation during the period involved. Acceptance of a temporary allocation reduction would not have any effect on the normal adjacency-based disbursements that a RAN would receive.

A possible basis for incentive payments would be to compensate according to the amount by which a RAN voluntarily reduced its allocation. Suppose, for example, that all RANs chose to take money rather than fish and froze their yellowfin catches at the "recent years" level given in Table 5. The overall reduction in their allocations (from the high allocation level) would amount to about 75,000 tons and would increase the unallocated share of the catch to 158,000 tons. How much would such a reduction mean to RANs in terms of increased participant fee disbursements? At $40 per ton for yellowfin, participant fee disbursements to non-RAN harvesting nations were $2.2 million in the recent years example of Table 5. If this entire amount were used to finance a system of incentive payments to RANs not harvesting their full allocations, they would receive $29.41 for each ton of unutilized allocation and total overall disbursements to RANs would increase from $5.4 million to $7.6 million. For a RAN without a fishery, receipts would just about double from $32.00 per ton to $61.41 per ton. But obviously not all RANs would want to freeze their catches at recent levels, and if less is frozen, payments to those who are willing to restrict catches could be increased. For example, if RANs agree to reduce allocations by only half the previous figure, or 37,500 tons, then those that did so could receive twice as large an incentive payment or $58.82 per ton of unutilized allocation. It would probably be desirable for per ton incentive payments to increase as allocation reductions decrease, because in marginal cases this would tend to discourage fleet development. But it might also be desirable to establish an upper limit on the per ton incentive beyond which payments would remain constant.

One reason remains for thinking that the impact of adjacency-based allocations on non-RANs might be considerably less severe than Figure 14 and Table 15 seem to suggest. Within the United States fishing industry there has been a trend for vessel ownership to be transferred from individuals to processing companies. This assures processors a reliable supply of raw fish, a consideration that is apparently important enough for them to be willing to buy into the fleet in spite of possible losses in the operation of their vessels. Presumably if vessel operating losses were to occur, they could be recovered in processing and merchandizing the final product, enabling overall profitability. Hence the trend toward processor ownership of vessels might provide something of a buffer enabling non-RANs to weather adverse effects associated with establishment of allocations.

Considering all of the mitigating factors mentioned so far, it appears that the impact of PAQ management on non-RANs would actually be considerably less severe than the fleet transition example suggests. (Recall that this example was intended to define the most extreme situation.) It is of obvious interest then to evaluate what the actual impact of PAQ management might be, even if this can only be done in a very speculative way. A straightforward approach to making such an evaluation is to take, as a point of departure, the examples of Tables 5 and 15 involving high allocations of both species and RAN fleet development by new construction. Starting from this most extreme situation for non-RANs, the effects of making various assumptions concerning mitigating factors can then be evaluated sequentially. Two possible situations are explored by this method. In the first situation (see Table 16, assumption 1) it is assumed that yellowfin and skipjack allocations are set equal to two-thirds of the average catches made by the entire international fleet in national 200-mile zones in recent years and that RAN fleets develop to harvest these allocations with the annual production of vessels entering RAN fleets being 3 tons per ton of carrying capacity. It is then assumed that 25 percent of these allocations remain unutilized by the RANs receiving them (assumption 2), perhaps in part as a result of an incentive program, and that fleets and catches are modified accordingly. Additional assumptions are that RAN fleet growth is 50 percent by flag changes from non-RANs and 50 percent by new construction (assumption 3); that RAN fleets are reduced by an additional 10 percent because of departures to other areas, obsolescence, and sinkings (assumption 4); and that non-RAN skipjack catches increase by 10 percent because of open access and perhaps because of improved fishing techniques (assumption 5). The second situation examined differs from the first in that only yellowfin are allocated; for this species it is assumed that there are high allocations equal the recent average catches by the entire fleet in national 200-mile zones (see Table 17, assumption 1). Assumptions 2–5 for the second situation are identical to those made in the first situation.

Tables 16 and 17 show the effect of each successive mitigating assumption on non-RANs in terms of increased catches, reduced fleet size, and increased catch per capacity ton. From the point of view of the individual boat owner the last factor, catch per capacity ton, is of key importance. Note that in the first situation with yellowfin and skipjack quotas set at two-thirds of their high levels (Table 16), the net effect of the first five assumptions is to more than double the annual non-RAN catch per capacity ton from 0.75 tons to 1.72 tons. At the same time the RAN catch per ton drops only slightly, from 2.56 tons to 2.24 tons. Results were quite similar in the second situation where there were high allocations of yellowfin only (Table 17), although in this case the first five assumptions lead to a somewhat smaller non-RAN catch per capacity ton of 1.58 tons. How do these results compare to the recent years (1970–74) situation in which there are no RAN allocations? To facilitate this comparison, figures on catches, fleet carrying capacities, and catches per capacity ton from the recent years example of Tables 5 and 15 have been incorporated into Tables 16 and 17. In each case they follow assumption 5 and appear in brackets. Note that in either allocation situation, after taking the first five assumptions into account, non-RAN catches and fleets are reduced as compared to the recent years example. Conversely, RAN catches and fleets are larger. The most interesting result, however, involves the annual catch per capacity ton for non-RANs. For both types of allocation this figure changes only insignificantly from the recent years example (either 1.72 tons or 1.58 tons versus 1.69 tons for recent years).

Although the results presented so far are admittedly rather speculative, they do tend to suggest that transition to a PAQ management system need not result in economic disaster for non-RAN fleets. Furthermore, there are two additional factors that slightly improve the actual economic outlook: fishing in the Atlantic and western Pacific after closure and catches of other species in the eastern Pacific. These factors are covered in the last two lines of Tables 16 and 17 where it is assumed that 10,000 tons of yellowfin and 10,000 tons of skipjack are taken annually after closure by non-RANs in the Atlantic and the western Pacific together and that 12,000 tons of bigeye and northern

TABLE 16. POSSIBLE SIGNIFICANCE OF FACTORS TENDING TO REDUCE THE IMPACT OF PAQ MANAGEMENT WITH RAN YELLOWFIN AND SKIPJACK ALLOCATIONS SET AT TWO-THIRDS OF THEIR HIGH LEVELS

| | Annual catches in the eastern Pacific (thousands of short tons) | | | | | | | | | Fleet carrying capacities (thousands of short tons) | | | Annual catch per capacity ton (short tons) | | |
| | RANs | | | Non-RANs | | | RANs and Non-RANs Combined | | | | | | | | |
	Yellowfin	Skipjack	Total	Yellowfin	Skipjack	Total	Yellowfin	Skipjack	Total	RANs	Non-RANs	Overall	RANs	Non-RANs	Overall
Starting Point High yellowfin and skipjack allocations with RAN fleet growth by new construction (see Table 5 and Table 15)	108.8	59.8	168.6	83.6	10.0	93.5	192.4	69.7	262.1	65.8	125.0	190.8	2.56	.75	1.37
Assumption #1 RAN allocations at two-thirds of the above high levels.	72.5	39.9	112.4	119.9	29.8	149.7	192.4	69.7	262.1	47.1	125.0	172.1	2.39	1.20	1.52
Assumption #2 One-quarter of the total RAN allocation is unutilized by the RANs and remains available to the non-RANs	54.4	29.9	84.3	138.0	39.8	177.8	192.4	69.7	262.1	37.7	125.0	162.7	2.24	1.42	1.61
Assumption #3 RAN fleet growth is one-half by flag changes from non-RANs and one-half by new construction.	54.4	29.9	84.3	138.0	39.8	177.8	192.4	69.7	262.1	37.7	119.4	157.1	2.24	1.49	1.67
Assumption #4 Ten percent of the remaining RAN fleet departs for other areas, becomes obsolescent, or sinks and is not replaced	54.4	29.9	84.3	138.0	39.8	177.8	192.4	69.7	262.1	37.7	107.5	145.2	2.24	1.65	1.81
Assumption #5 Non-RANs skipjack catches increase by ten percent due to open access and/or development of better fishing methods.	54.4	29.9	84.3	138.0	46.8	184.8	192.4	76.7	269.1	37.7	107.5	145.2	2.24	1.72	1.85
(Figures in parentheses are from the recent years examples of Table 5 and Table 15 and are for comparison.)			(50.5)			(211.6)			(262.1)	(26.4)	(125.0)	(151.4)	(1.91)	(1.69)	(1.73)
Assumption #6 Non-RANs take 10,000 tons each of yellowfin and skipjack from the Atlantic and western Pacific.	54.4	29.9	84.3	148.0	56.8	204.8	202.4	86.7	289.1	37.7	107.5	145.2	2.24	1.91	1.99
Assumption #7 Catches of bigeye and bluefin in the eastern Pacific amount to 12,000 tons.			88.0			213.1			301.1	37.7	107.5	145.2	2.33	1.98	2.07

NOTE: Mitigating factors are considered sequentially starting from the example of Table 5 and Table 15 in which there are high yellowfin and skipjack allocations and RAN fleet growth is by new construction.

TABLE 17. POSSIBLE SIGNIFICANCE OF FACTORS TENDING TO REDUCE THE IMPACT OF PAQ MANAGEMENT WITH RAN YELLOWFIN ALLOCATIONS SET AT THEIR HIGH LEVELS AND SKIPJACK UNALLOCATED

| | Annual catches in the eastern Pacific (thousands of short tons) | | | | | | | | | Fleet carrying capacities (thousands of short tons) | | | Annual catch per capacity ton (short tons) | | |
| | RANs | | | Non-RANs | | | RANs and Non-RANs Combined | | | | | | | | |
	Yellowfin	Skipjack	Total	Yellowfin	Skipjack	Total	Yellowfin	Skipjack	Total	RANs	Non-RANs	Overall	RANs	Non-RANs	Overall
Starting Point High yellowfin and skipjack allocations with RAN fleet growth by new construction (see Table 5 and Table 15)	108.8	59.7	168.5	83.6	10.0	93.6	192.4	69.7	262.1	65.8	125.0	190.8	2.56	.75	1.37
Assumption # 1 High yellowfin allocations as above, but no skipjack allocations.	108.8	27.7	136.5	83.6	42.0	125.6	192.4	69.7	262.1	55.0	125.0	180.1	2.48	1.00	1.46
Assumption # 2 One-quarter of the total RAN allocation is unutilized by the RANs and remains available to the non-RANs.	81.6	24.9	106.5	110.8	44.8	155.6	192.4	69.7	262.1	47.9	125.0	172.9	2.22	1.24	1.52
Assumption # 3 RAN fleet growth is one-half by flag changes from non-RANs and one-half by new construction.	81.6	24.9	106.5	110.8	44.8	155.6	192.4	69.7	262.1	47.9	114.2	162.1	2.22	1.36	1.62
Assumption # 4 Ten percent of the remaining RAN fleet departs for other areas, becomes obsolescent, or sinks and is not replaced.	81.6	24.9	106.5	110.8	44.8	155.6	192.4	69.7	262.1	47.9	102.8	150.7	2.22	1.51	1.74
Assumption # 5 Non-RANs skipjack catches increase by ten percent due to open access and/or development of better fishing methods.	81.6	24.9	106.5	110.8	51.8	162.6	192.4	76.7	269.1	47.9	102.8	150.7	2.22	1.58	1.79
(Figures in parentheses are from the recent years example of Table 5 and Table 15 and are for comparison.)			(50.5)			(211.6)			(262.1)	(26.4)	(125.0)	(151.4)	(1.91)	(1.69)	(1.73)
Assumption # 6 Non-RANs take 10,000 tons each of yellowfin and skipjack from the Atlantic and western Pacific.	81.6	24.9	106.5	120.8	61.8	182.6	202.4	86.7	289.1	47.9	102.8	150.7	2.22	1.78	1.92
Assumption # 7 Catches of bigeye and bluefin in the eastern Pacific amount to 12,000 tons.			111.3			189.8			301.1	47.9	102.8	150.7	2.32	1.85	2.00

NOTE: Mitigating factors are considered sequentially starting from the example of Table 5 and Table 15 in which there are high yellowfin and skipjack allocations and RAN fleet growth is by new construction.

bluefin are available annually in the eastern Pacific. While these examples provide a basis for optimism concerning the likelihood of successful PAQ management, the optimism is only justified if involved parties are willing to negotiate in good faith toward a solution that recognizes the legitimate needs and aspirations of all.

Appendix IV.
Control of Entry

A serious problem of excess capacity exists in many of the world's tuna fisheries, and this problem must be effectively addressed by any management system that is to be viable. If a rational means cannot be developed for maintaining fleets at some reasonable size, then economic and political pressures will almost certainly develop to overthrow any system for determining yield levels, allocating catches, and enforcing regulations. A possible means for controlling fleet growth in a PAQ management system is described in this appendix in some detail, primarily to illustrate the complexity of the problem and to emphasize the importance of developing a logical approach to it. As usual, concepts will be developed with the specific example of the eastern Pacific yellowfin and skipjack fishery in mind. Excess carrying capacity is an especially serious problem in this fishery.

An Illustrative Licensing System for Control of Fleet Growth

It can be assumed that any plan for controlling fleet size will involve establishment of some sort of system for issuance of licenses allowing vessels to fish, and anyone or any body attempting to design such a system must do two things. First, a noncontradictory set of requirements and objectives must be established and explicitly spelled out. Second, a set of procedures for implementing and operating the system must be established that is fully consistent with the stated requirements and objectives. Included would be procedures governing each possible type of license transition that could take place in the fleet. Unless real care is taken to insure that all procedures are consistent with the explicitly stated goals, it is almost certain that there will be inequities and inconsistencies. But if everything is carefully formulated initially, future disputes and misunderstandings can hopefully be avoided. A possible system for control of fleet size will now be described. While it will probably seem complicated, this reflects the complexity of the real life problem, and should serve as a caution against seeking quick and easy solutions. Careful thinking should precede discussions on matters as complicated as control of fleet size.

Before formulating a set of requirements and objectives, it is convenient to distinguish between two types of RANs. First, there are RANs with fleets that are underdeveloped in the sense that their recent catches are less than their allocations. Such nations will be referred to as URANs. Second, there are RANs with fleets that are well developed in the sense that their recent catches equal or exceed their allocations. These nations will be referred to as DRANs. Of course, non-RANs will also be considered.

Five requirements and objectives that might seem reasonable to many can now be stated in very simple terms:

1. In implementing a system for control of fleet size, all vessels in the existing fleet should be subject to the same set of rules, and none should be arbitrarily excluded from the fishery when a system is adopted.
2. A maximum total fleet size should be established.
3. Establishment of a maximum fleet size should not interfere with development of any URAN fleet up to the point where it is capable of harvesting its full allocation.
4. Non-RAN and DRAN vessels should be treated on an equal basis, including those of non-RANs not initially participating in the fishery.
5. When an existing fleet is already too large (as it is in the eastern Pacific) policies should be adopted that tend to reduce maximum fleet size whenever possible.

To meet the first requirement, all vessels presently participating in the fishery could be issued licenses to fish. Because of significant differences in carrying capacity, type of gear used, age of vessel, and other factors, licenses should probably be issued in terms of standardized units of gear. For example, in the eastern Pacific a standardized unit of gear might be 100 tons of carrying capacity on a modern purse seiner in the 1,000–1,100 ton class. Such a vessel would be issued a license for 10 gear units. Larger and smaller vessels would receive licenses for more or fewer standardized gear units. Factors other than carrying capacity could also be considered, such as type of gear and age of vessel. In any event, the objective would be for each vessel's license to realistically reflect its true fishing efficiency. License holders could be individuals, corporations, or nations.

Once the licenses were issued, there would be specified procedures for transfers from license holders to license seekers. It could be left up to the holders to negotiate transactions with those seeking licenses, or the international management agency could provide a license brokerage service, or some other arrangement could be made. The key point is that, with the exception of those involved in developing URAN fleets, all license seekers should have an equal opportunity to acquire them in accordance with requirement 4 of the above list.

The second requirement was that a maximum fleet size should be established. There are a number of ways to accomplish this. One logical approach would be to start with the present fleet and allow for sufficient growth of URAN fleets to enable them to fully utilize their allocations as mandated by the third requirement. Suppose that each URAN fleet is allowed to develop until the added vessels are collectively harvesting the difference between the URAN's allocation and its present catch at an annual rate of three tons per standardized ton of carrying capacity. Under this criterion the maximum standardized size of the international fleet in terms of carrying capacity would be:

$$
\left(\begin{array}{c} \text{Present standardized} \\ \text{capacity of the} \\ \text{international fleet} \end{array} \right) + \frac{\left(\begin{array}{c} \text{Total URAN} \\ \text{allocation} \end{array} \right) - \left(\begin{array}{c} \text{Present URAN} \\ \text{catch} \end{array} \right)}{3}
$$

In the eastern Pacific, if only yellowfin are allocated, then the above expression could and should be modified to include catches of skipjack. Cases 1–8 of Table 14, suggest how this might be done. See also Table 15, Figure 14, and the accompanying text.

URAN fleet growth can occur in several ways. Some involve addition of vessels to the fleet (e.g., new construction), while others do not (e.g., flag change from a non-RAN fleet to a URAN fleet). In the latter case, an opportunity exists for reducing maximum fleet size in keeping with the fifth objective without compromising any of the other objectives. A pool of licenses could be created at the outset for issuance to vessels which are entering both a URAN fleet and the eastern Pacific fishery. Initially this pool could be large enough to cover all URAN fleet growth. But if a vessel presently participating in the fishery transferred into a URAN fleet from a non-RAN or DRAN fleet, its license could transfer with it, and both the pool of licenses reserved for URAN fleet development and the maximum potential fleet size could be reduced by an appropriate amount. Such an arrangement would be fully consistent with requirement 3 providing for URAN fleet development.

The measures outlined so far suggest in general terms how a program to limit fleet size could be initiated within the framework of the five requirements and objectives mentioned earlier. It remains to establish specific rules for handling each of the many types of changes that could occur in the fleet. Thirty-three types of transactions are possible as indicated in Table 18. Because DRANS and non-RANs are treated the same in all transactions, and because it makes no difference for present purposes how a vessel enters or departs from the regional fleet, these 33 transactions can be grouped into ten categories. The exact manner for handling transactions falling into each category is indicated in Table 19. Immediate effects on URAN, DRAN, and non-RAN fleets are indicated, as are effects on fleets of individual nations involved in transactions. In addition, license transactions associated with each

TABLE 18

33 POSSIBLE TYPES OF CHANGES, GROUPED IN 10 CATEGORIES (A-J), IN REGIONAL URAN, DRAN, AND NON-RAN FLEETS

Additions to Regional Fleet

Method of entry	Fleet to which capacity is added		
	URAN	DRAN	Non-RAN
New construction	B	A	A
Interregional transfer with flag change	B	A	A
Interregional transfer without flag change	B	A	A

Departures from Regional Fleet

Method of departure	Fleet from which capacity departs		
	URAN	DRAN	Non-RAN
Retirement of vessel	D	C	C
Loss of vessel	D	C	C
Interregional transfer with flag change	D	C	C
Interregional transfer without flag change	D	C	C

Intraregional Transfers between Nations (Flag Change)

Fleet from which capacity transfers	Fleet to which capacity transfers		
	URAN	DRAN	Non-RAN
URAN	H	G	G
DRAN	F	E	E
Non-RAN	F	E	E

Intraregional Transfers within a Nation (No Flag Change)

	Fleet within which capacity transfers		
	URAN	DRAN	Non-RAN
	J	I	I

TABLE 19. EFFECTS OF 10 TYPES OF FLEET CHANGE (A-J) ON URAN, DRAN,

Type of change	Fleet components involved in change	Category of change
Additions to regional fleet (new construction, inter-regional transfers with or without flag changes)	New entrant to DRAN or non-RAN fleet	A
	New entrant to URAN fleet	B
Elimination from regional fleet (sinkings, retirements, inter-regional transfers with or without flag changes)	Departure from DRAN or non-RAN fleet	C
	Departure from URAN fleet	D
Intra-regional transfers involving flag changes	Within combined DRAN and non-RAN fleet	E
	From DRAN or non-RAN fleet to URAN fleet	F
	From URAN fleet to DRAN or non-RAN fleet	G
	Within URAN fleet	H
Intra-regional transfers not involving flag changes	Within DRAN or non-RAN fleet	I
	Within URAN fleet	J

Type of change	Fleet components involved in change	Category of change
Additions to regional fleet (new construction, inter-regional transfers with or without flag changes)	New entrant to DRAN or non-RAN fleet	A
	New entrant to URAN fleet	B
Elimination from regional fleet (sinkings, retirements, inter-regional transfers with or without flag changes)	Departure from DRAN or non-RAN fleet	C
	Departure from URAN fleet	D
Intra-regional transfers involving flag changes	Within combined DRAN and non-RAN fleet	E
	From DRAN or non-RAN fleet to URAN fleet	F
	From URAN fleet to DRAN or non-RAN fleet	G
	Within URAN fleet	H
Intra-regional transfers not involving flag changes	Within DRAN or non-RAN fleet	I
	Within URAN fleet	J

NOTE: The 10 categories of fleet change are defined in Table 18. In sections dealing with effects of change, a plus (+) indicates an increase, a minus (−) indicates a reduction, and a zero (0) indicates

Immediate effect of change on fleet size

Effect on major fleet components			Effect on fleets of specific nations involved in change	
URAN	DRAN and non-RAN	Entire fleet	URAN	DRAN or non-RAN
0	+	+		+
+	0	+	+	
0	−	−		−
−	0	−	−	
0	0	0		+ and −
+	−	0	+	−
−	+	0	−	+
0	0	0	+ and −	
0	0	0		0
0	0	0	0	

Licensing transactions associated with each category of change

License transfers with vessel	License issued from URAN pool	Change in pool of licenses reserved for URAN fleet development		License for a DRAN or non-RAN --		Effect of change on maximum potential fleet size
		Reduced	Increased	Becomes Available	Must be acquired	
					*	0
	*	*				0
				*		0
			*			0
*						0
*		*				−
			*		*	0
*						0
*						0
*						0

no change. In sections dealing with licensing transactions, an asterisk (*) means the indicated action is required or takes place.

category of change are shown. In some transactions licenses transfer with the vessel. In others a new license is issued from the pool reserved for URAN fleet development. Finally, some require acquisition of a license from a vessel departing from the fishery for one reason or another. For each licensing transaction the long-term effect on maximum potential fleet size is also indicated.

Tables 18 and 19 must be studied together carefully in order to appreciate the consistency of the proposed system with the five objectives listed earlier. While the motivation for most entries in Table 19 is obvious, some require further explanation. Categories A through D deal with entry and departure of vessels into and out of the region's fleet considered as a whole. Before a new-entrant vessel can join a DRAN or non-RAN fleet (category A), it must first acquire a license from a comparable vessel that is leaving the fishery (category C). This is consistent with the third requirement that there should be a clearly defined maximum fleet size. New entrants joining URAN fleets (category B) need not acquire a license from a departing vessel; instead they receive licenses from the pool reserved for URAN fleet development. The case of a URAN vessel departing from the fishery (category D) is a little more complicated. If it was allowed to transfer its license to a DRAN or non-RAN vessel, and if the URAN is subsequently to be allowed to replace the vessel by drawing a new license from the URAN pool, the end result would be that the URAN pool would be exhausted before all URAN fleets were fully developed, in violation of the third requirement. If additional licenses were added to the URAN pool to avoid this problem, then the second requirement to limit fleet size would be violated. To avoid this dilemma, when URAN vessels leave the fishery their licenses must go into the URAN pool where they remain available until departing vessels are replaced. Note that if a vessel is retired or sinks, a URAN, DRAN, or non-RAN owner should have the option of retaining his license and replacing the vessel.

Intraregional transfers involving flag changes are covered in categories E through H. Categories E and H are simple transfers not resulting in growth or depletion of the overall URAN fleet. Category F is potentially very important. It covers flag changes from DRANs or non-RANs to URANs, and in the eastern Pacific this would primarily mean flag changes from the United States fleet to Latin American fleets. In these category F flag changes, licenses transfer with vessels and equivalent numbers of licenses are eliminated from the pool reserved for URAN fleet development without being actually issued. Hence category F transactions reduce maximum potential fleet size in keeping with the fifth objective. Category G flag changes from URAN to DRAN or non-RAN fleets would be rare but still require some explanation. If the URAN is to replace the departing vessel without either depleting the URAN license pool or increasing maximum fleet size, the license of the departing vessel must go into the URAN license pool rather than with the vessel. (Categories G and

D are similar in this respect.) Thus the DRAN or non-RAN acquiring the vessel must first acquire a license from a departing DRAN or non-RAN vessel. This treatment is consistent with all of the stated requirements. The last two categories, I and J, cover simple within-category transfers and require no comment.

When a developing URAN fleet becomes large enough for the URAN to be reclassified as a DRAN, it would no longer draw licenses from the pool reserved for URAN fleet development. However, it could later revert to URAN status and once again utilize the pool. For example, taking all figures to be in standardized tons of carrying capacity, suppose a DRAN is 500 tons over the dividing line and decides to transfer 1,200 tons to a non-RAN. This can be thought of as a combination of a category E and a category G transaction. The license for 500 tons would transfer with the vessel (category E). The license for the remaining 700 tons would be added to the URAN pool, and the non-RAN would have to acquire from elsewhere a license covering this 700 tons (category G). Next suppose the same nation, now classified as a URAN, decides to build a new 1,000 ton seiner. A license for 700 tons would be drawn from the URAN pool (category B) putting the nation back into the DRAN category, and a license for the remaining 300 tons would have to be acquired (category A).

This completes the detailed description of one possible licensing system for controlling fleet size. The primary reasons for going into this degree of detail have been: (1) to illustrate the importance of developing licensing procedures that are consistent with a clearly stated set of requirements and objectives; and (2) to emphasize that the problem of controlling fleet size is inherently complex and not to be treated superficially. In addition to licensing there are a number of other important issues pertaining to control of fleet size that must be considered. Some of these will be briefly touched on now.

Further Considerations and Problems

Under the system just described, the transferable licenses that would be created and issued to all present participants and to future category B entrants into URAN fleets could have considerable property value, depending on the handling of transfers. For example, if a prospective entrant had to bid against other prospective entrants to acquire an appropriate license, licenses would likely become very valuable to those holding them. This raises an important question. If licenses were issued initially without charge, their subsequent value would represent windfall benefits to their recipients provided they had the option of selling their licenses. Some might deem this only just, considering all of the management restraints that would be established concurrently, and point out that such benefits provide an incentive for adoption of a licensing system. Others might feel that windfall benefits are improper. If the latter view were to prevail, when licenses were first issued fees approximating their current market value might be levied.

A related complication arises from the differences in the ways that URANS on the one hand and DRANS and non-RANS on the other dispose of vessels (see Table 18). These differences could result in a URAN license being less valuable to its owner than a comparable DRAN or non-RAN license. A URAN owner could transfer his permit to another URAN owner only through category H or J transfers, but in other transactions (categories D and G) licenses would return to the URAN pool. In contrast, DRAN and non-RAN owners would realize the value of their licenses no matter how they disposed of their vessels (categories C, E, F, and I). Clearly the market for URAN held licenses would tend to be weaker than that for DRAN or non-RAN held licenses. Differential license values could generate long-range concern that prospective URAN owners might tend to favor category B (free licenses) or category H and J (lower-priced licenses) transactions that would not reduce overall fleet size in preference to category F (higher-priced licenses) flag changes that would reduce maximum potential fleet size.

To avoid (or at least minimize) problems relating to the value of licenses, nations might prefer that all license transactions be handled through the international management body rather than negotiated by the owners themselves. For example, all license transfers might be processed on a fixed fee basis with proceeds going to the owner. In this case priority rules would have to be established dictating which of several applicants should get an available license. Alternatively, transfer fees could be determined by bidding, perhaps with a fixed minimum bid going to the former owner and anything above the minimum going to the agency. Even simpler would be for the agency to keep all proceeds, whether derived from fixed fees or bids, thus entirely eliminating the property value of the license to the owner.

In the licensing system under consideration, potential fleet size is reduced by reducing the pool of licenses reserved for URAN fleet development whenever there is a category F flag change from a DRAN or a non-RAN to a URAN fleet. However, this is the only provision relevant to objective 5, the reduction of potential fleet size. Additional provisions aimed at achieving this goal are possible. For example, URAN fleet development by category B transactions (vessels that enter URAN fleets are also new entrants into the fishery) could be partially restricted or even prohibited entirely. If such restrictions were agreed to, the pool of licenses reserved for URAN fleet development would be reduced by an appropriate amount or, in the case of total prohibition, eliminated entirely. If the pool were eliminated entirely, all category B and D transactions (new entries and departures) would be handled like category A and C transactions, respectively, and all intraregional transfers, categories E through J, would be transacted in the same way (licenses transferring with vessels). In effect, the distinction between a URAN on the one hand and a

DRAN or non-RAN on the other hand would be eliminated. Although at best this approach would limit the eastern Pacific fleet to its present size (probably already too large), it could be considered a step in the right direction. However, while URAN fleet growth would not be prevented, this arrangement could be construed as not in keeping with the second objective, that gear limitation should not interfere with URAN fleet growth.

If reducing fleet size is considered a more realistic goal than merely limiting fleet expansion, then an obvious approach would be to only partially replace departing DRAN and non-RAN vessels (category C) until some optimum or agreed upon fleet size was reached. (Perhaps any fleet size that could be agreed upon should be considered as "optimum" in a certain sense!) The rate at which the desired fleet size was approached would depend both on how rapidly vessels departed from the fleet and on the replacement ratio. For example, 1,000 tons could be replaced by 0 tons, 100 tons, 200 tons, and so on, and licenses for the remaining tonnage eliminated, perhaps with some compensation to its holder. Under such a partial replacement scheme, a desirable exception might be to allow full replacement of vessels that sink or become too old to use. Also, because a partial replacement policy should not interfere with URAN fleet growth, URAN departures (category D) could be fully rather than partially replaced.

The possibility of URANs voluntarily refraining from developing their fleets in exchange for increased participant fee disbursements was discussed in Appendix III within the context of minimizing the impact of adoption of a PAQ management system on non-RAN catches. Such voluntary restraints are also germane to the problem of controlling fleet size. To the extent that URANs voluntarily refrain from developing their fleets, the license pool for URAN fleet development could also be temporarily reduced. But because URANs might all eventually decide to develop their fleets fully, there would be no permanent reduction of the maximum potential fleet size.

In discussing fleet limitation it has been tacitly assumed that RAN allocations would be nontransferable guarantees of access. This assumption, for example, underlies the idea of establishing a pool of licenses for URAN fleet development based on differences between allocations and recent catches. If, instead, allocations were freely transferable property rights, would any modifications of the proposed licensing system be necessary or desirable? For example, would it be reasonable to allow a non-RAN that acquires part of a URAN quota to draw licenses out of the URAN license pool? Or should the URAN license pool be reduced when an allocation is transferred? The answer in both cases is: probably not. The decision to acquire a vessel is, in general, a decision with fairly long-term consequences. On the other hand, allocation transfers could be made on any basis ranging from annually to long term. At some time in the future a URAN might want to transfer its allocation to another party or build up its own fleet to utilize its allocation. Modification of the gear limitation system seems inappropriate because of these uncertainties. If allocations are transferable and carried out independent of the vessel licensing system, a URAN could transfer its allocation and also develop its fleet. However, such a policy would be unwieldy, illogical, and unlikely to be adopted. All in all, vessel licensing seems somewhat incompatible with the concept of transferable allocations, an argument for treating allocations as guarantees of access instead.

Thus far the discussion of specific controls on how licenses would be issued, who would receive them, and how they would be transferred has been predicated on the belief that some sort of overall control of fleet size is essential to the establishment and continuing viability of an international management program. But RANs (both URANs and DRANs), because they receive national allocations, might feel that it should be strictly up to themselves as individual nations to decide how to utilize their allocations and determine their fleet size. For example, one RAN in seeking maximum profits might prefer a fleet no larger than the minimum size necessary to harvest its full allocation; another RAN pursuing a social goal such as maximum employment might want to take its allocation with as large a fleet as possible. Interestingly, a possible means for accommodating both of these divergent objectives exists within the context of a system for control of fleet size, and it has already been discussed in Appendix I, which deals with closure of the fishery.

The basis of the closure dilemma was whether RAN catches should count against allocations from the beginning of the open fishing season or only after closure. This dilemma, it was suggested, could be resolved by splitting national fleets into two components. One component, the allocation fleet, would havest the entire national allocation and could fish after closure if necessary. A second component, the open-season fleet, would compete on equal footing with all other vessels engaged in harvesting the unallocated portion of the resource and would stop fishing when the season closed. Under such an arrangement a RAN could make its allocation fleet as large or small as it desired, and widely divergent national goals could be readily accommodated.

Without going into great detail, it can be seen how a split-fleet solution to the problems of closure and divergent RAN goals could be compatible with control of fleet size. Because each RAN would be responsible for determining the size of its own allocation fleet, it would only be necessary to control the size of the open-season fleet. All open-season vessels would be treated similarly and distinctions between URANs, DRANs, and non-RANs would be unnecessary. Initially, each nation would designate individual vessels as belonging to either its allocation fleet or its open-season fleet, and transferable licenses would be issued only to vessels in the open-season fleet. Thereafter, any new entrants to the open-season fleet, including vessels transferring from a nation's allocation fleet, would first have to acquire a license.

If there was a split-fleet fishery in the eastern Pacific, the open-season fleet would include all non-RAN fleets and would initially be large compared to available resources. In order to reduce the open-season fleet to a more desirable size, it would be reasonable to replace departing vessels only partially. A complication would arise, however, in the case of transfers from the open-season fleet to developing RAN allocation fleets. Such transactions would be similar to category F flag changes in that vessels continue to fish in the eastern Pacific. Also, they would be similar to category C flag changes in that licenses would not transfer with vessels. But category F changes reduce potential fleet size, while category C changes result in either unchanged or reduced potential fleet size depending on the replacement policy (see Table 19). In contrast, with a split fleet, transfers into allocation fleets would result in overall fleet growth under a partial replacement policy. Hence, even though the open-season fleet would be reduced, when a vessel departed to join a developing allocation fleet (through a flag change or by being redesignated by the RAN), its license should probably be completely retired. Other departures, including departures to allocation fleets of RANs already taking their full allocations, should be on a partial replacement basis until the desired open-season fleet size is achieved. An exception to these policies would be made in the case of a RAN exchanging equivalent capacity between its two fleets in which case the license would simply be transferred.

Thus far the problem of controlling fleet size has been considered from essentially a regional point of view, and it has been assumed that the objective is to reduce fleets or at least curtail fleet growth. But if a regional fishery is underdeveloped in the sense that fleets are small compared to the apparent resources, then further growth is indicated. It would obviously be desirable to transfer vessels from overdeveloped fisheries into an underdeveloped fishery of this type. Exploration efforts by United States purse seiners in western Pacific waters near New Zealand could foreshadow exactly such a transfer. In areas with underdeveloped fisheries it would be advisable to establish programs for control of fleet growth before fleet size becomes a problem. If this could be accomplished, major problems plaguing overdeveloped tuna fisheries could be averted with important economic and resource management benefits. Ideally regional fleet growth could be controlled on a worldwide basis. With proper coordination through an international body, it would be a straightforward task to identify opportunities for transferring vessels from overdeveloped to underdeveloped fisheries. In fact, substantial incentives such as free licenses might be provided to encourage desirable interregional transfers.

Index

Africa, 111, 112

Albacore *(Thunnus alalunga):* migrations, 5; world catch, 6; catch by ocean, 7; mentioned, 89, 194, 223

Allocations: based on resource adjacency, 14, 19, 120; based on average catch, 61; none beyond 200 miles, 61; no skipjack quota, 61; transfer of, as property rights, 62, 84–85; nontransferable guarantees of access, 62, 84, 198; unutilized catch harvested by non-RAN fleet, 62, 200; modification of participant fee system, 84–85; RAN *de facto* catch limit, 86–87; two-tier national fleets, 86–87; economic factors, 120; total catch allocation at international level, 120; of commercially important species in northwest Atlantic, 122–23; competitive bidding system, 135; fixed participant fees auctioned, 135; unutilized catches auctioned, 135; international agency to distribute funds, 135; and fleet transition problems, 207; increase in skipjack and tuna catch related to fleet growth, 215; incentive payments for RAN unutilized allocations, 223–24; per ton payments increase with decreased allocations, 223–24; mentioned, 60, 73. *See also* PAQ management system

American Tuna Boat Association, 145

Angola, 18

Antarctic Ocean, 122

Atlantic Ocean: fished by tuna seiners after CYRA closure, 111; northwest fisheries, 120–23; total international catch allocations, 120–23; regional fisheries management, 193; mentioned, 89–90, 112, 122, 197, 205, 222, 225, 228. *See also* ICCAT; ICNAF

Australia, 174, 176

Bait boats: licensing of, 73; tuna longline fisheries, 137; Ecuadorian fleet, 204

Bait fish, 194–95

Bait fishing: basic technique, 8–9; unsuccessful in eastern Pacific, 154; mentioned, 136

Bering Sea, 121

Bermuda: participant in eastern Pacific tuna fishery, 27, 112; big game fishing, 176

Bigeye tuna *(Thunnus obesus):* migrations, 5; world catch, 6; catch by ocean, 7; mentioned, 62, 78, 89, 194, 205, 223, 225–28 *passim*

Big game fishing: major centers, 176–77

Billfish: definition of, 5; world catches, 6, 176; longline catches, 62; taxonomy, 173; species of, 173; migrations, 173–74; tag returns, 173–74; commercial versus sport fishing, 173–79 *passim;* behavior of, 174; taken with longline gear, 174; markets for, 174; annual value of, 176; sport fishing for big game fishing centers, 176–78; recreational value, 176–78; management needed, 177–79; Japanese longline fishery off Mexico, 178; sport fish sanctuaries, 179; broad research program needed, 179; mentioned, 194, 205

Birds: fish schools located by sightings of, 138

Bluefin tuna: migrations, 5; world catch, 6; catch by ocean, 7; catch limited, 35; distribution of stocks, 193; mentioned, 62, 78, 89, 173, 181, 194, 205, 223, 225, 228

Bonito, 62, 89, 194, 223

Brazil, 18, 173–74

Cabo Blanco, Peru, 176

California, 208, 209

Canada: IATTC member, 13; ICCAT member, 18; harvesting nation in eastern Pacific tuna fishery, 27, 103, 108; fleet fishes Atlantic after CYRA closure, 111; historical participant in eastern Atlantic fishery, 112; northern fur seals, 121; mentioned, 94, 110, 114, 173
Carrying capacity. *See* Fleet carrying capacity
Catches: by species, 5, 6, 29, 30–31, 70–71, 80–81, 118–19, 213; by ocean, 7; by nation, 9, 11, 30–31; possibilities of increased production, 11–12; value of,15; fluctuations in yellowfin and skipjack catch, 29; yellowfin and skipjack taken in eastern Pacific, 1961–74, 29; annual per ton capacity taken by RANs and by non-RANs, 90; billfish, 176; RAN allocations allow maximum catches, 198
Catch per unit of effort: yellowfin, 118; skipjack, 119
Catch tax: participant fee system, 50. *See also* Participant fees
Central America: tuna-porpoise associated fishery, 137; mentioned, 49, 68, 167
Chile: historical participation in tuna fishery, 103, 108; skipjack resources, 108; mentioned, 94, 110, 114,
China, Republic of: major harvester of world catch of tuna, 26–27; longline fishing, 33–34, 154; catches dropped, 33–34; effect of total allocation system on, 130
Clipperton Island, 27, 64–65, 94
Closure: PAQ management system procedures, 62, 86–87; economic and philosophic conflicts, 199–200; effect on RANs, 199–201
Coalitions: between RANs only, 93–95; between RANs and non-RANs, 93–95; based on resource supply and fleet size, 93–95; effect on tuna resources, 94–95; advantages, 95–96; effect on fisheries beyond 200 miles, 96–97; effect on spawner-recruit (S-R) yellowfin relationship, 96–97; closed regional coalitions, 109; open to all nations, 113
Colombia, 27, 63, 213
Commercial fishermen: tuna, 136; billfish, 173; conflicts with sport fishermen, 173

Commission's Yellowfin Regulatory Area. *See* CYRA
Common property resources, 60, 77, 199
Competitive bidding system: structure, funding, administration, 131–35 *passim;* advantages and disadvantages, 131–35 *passim;* international tuna management organization mandatory, 131–32; harvesting rights assigned by bidding, 131–32; bid payments, 131–32; bids open to individuals, organizations, or nations, 133; speculation control, 133; payment guarantees, 133; licenses, 134; unfair to RANs, 134; allocations and methods of closure, 135; multispecies management for yellowfin and skipjack, 135; control of fleet size, 135
Congo, 27, 112
Conservation of resources: defined, 40; tuna, 42
Consumption of tuna: by nation, 11
Costa Rica: IATTC member, 13; jurisdiction over fisheries, 22; yellowfin and skipjack catch areas, 29; fleet expansion, 46; licenses required within 200-mile zone, 54–55; yellowfin allocations, 63, 80–81, 83; coalition combinations, 94; IATTC proposed allocation of yellowfin catch in CYRA, 127; mentioned, 27, 49, 64–67 *passim,* 74, 96, 103–4, 192, 212
Cuba, 18
CYRA (Commission's Yellowfin Regulatory Area): IATTC sets quotas and monitors fleet catches in, 36; vessels fishing west of, unrestricted, 36, 53–55 *passim;* fishing rights in 200-mile zone, 97; simulation model for yellowfin fishery, 97; boundary, 140; closure of fishery, 153; mentioned, 55, 111, 125, 126, 138

Disbursements: of participant fees based on 1970–74 catch in eastern Pacific, 82; under PAQ management system, 83
Dolphin. *See* Porpoise

Eastern spinner porpoise, 140, 150–51
Economic influence zone: extension to 200 miles, 72
Ecuador: claims 200-mile territorial seas, 22; yellowfin and skipjack catch areas,

29, 46, 204–5; fishing fleet, 46, 47, 199, 204–5; high license fees, 49, 79, 204–5; license required to fish within 200 miles of coast, 54–55; yellowfin allocations, 63, 66, 83, 199; skipjack resources, 67, 87–88, 199, 204–5; skipjack management under PAQ system, 87–88; coalition possibilities, 94–95; economic effect of open-access skipjack fishery, 204–5; bait boat fleet, 204–5; foreign vessels must fish 60 miles from shore, 204–5; PAQ management closes foreign vessels to 12-mile zone, 204–5; mentioned, 27, 64–65, 68, 103, 114, 115, 127, 213

Elizabeth, C.J, 164, 166–67

El Salvador, 27, 64–65

Enforcement: in eastern Pacific, 17–18; in Atlantic, 19–20; need for international regulations to conserve tuna resources, 35–36; and effective catch inspection, 35–36; difficulties of, 50; for licensing systems, 50; and cost of, 50; need for severe penalties, 50; basic to international management systems, 91–92; no agency now responsible, 91–92; surveillance system needed to inspect landings, 91–92; vessel monitoring by satellite possible, 91–92

Exclusive fishing zones: unacceptable approach to tuna management, 47

Exploitation of fishery: success measured by magnitude of catch, 60

Fisheries conservation. *See* Fisheries management

Fisheries management: basic requirements, 23; requires real-time monitoring as basis for effective decisions, 23; and statistical data on fishing effort by species, 23; and life history studies and environmental relationships, 23; minimum level of conservation defined, 41; objectives of IATTC, 41; regulation needed beyond 200-mile zone, 44; inadequate under present systems, 93, 180; national zones individually control overfishing, 94; simulated data models show depletion of stocks without effective regulation, 100, 104; and potential economic loss, 104; of tuna and associated species, criteria for, 180; international political support lacking,

180; problems of stock migration, fleet mobility, inadequate data, 181; single international organization needed for entire range of stocks, 181; responsibilities, objectives, implementation, and functions of global system, 181, 183; global organization system design, 181–97 *passim;* RAN claims, 199; effect on fishery of refusal to observe closure dates, 201; utilization of tuna resources, 202–5; international fleet composition, 206–28 *passim*

Fishing effort: data lacking in western Pacific, 23

Fishing methods and gear: for tuna-porpoise associated schools, 139; new porpoise-saving gear and fishing techniques needed, 154–55; must benefit fisherman economically, 154–55; long-term gear development program needed, 154–55; backing down technique to release porpoise, 162; raft man and speed boat operators reduce porpoise mortality, 164–66. *See also* Bait fishing; Longlining; Purse seining

Fishing mortality rates: age specific, 97

Flag changes. *See* Fleet development; Vessels

Fleet carrying capacity: world, 11; in eastern Pacific, 15–17; in Atlantic, 19; rapid growth in, 32–34; international controls needed, 32–34; controls complicated by vessel mobility, 32–34; vessels of RANs mostly inadequate to harvest resources, 66; flag transfers, effect on RANs and non-RANs, 89; licenses based on standardized vessel capacity, 90–91; replacement of vessels, 90–91; and PAQ management system, 90–91, 215–22 *passim;* large RAN fleets economically unsound, 199; projected changes over 10-year allocation phase-in period, 216–22 *passim*

Fleet composition: allocation fleets, 200; open-season fleets, 200; changes in international, 206–23 *passim;* vessel carrying capacity changes over 10-year allocation phase-in period, 218–22; U.S. vessel transfers from individual to processor ownership, 224; evaluation of PAQ impact on, 224–25; catch per capacity ton, 225

Fleet control: excess carrying capacity, 229; system for licensing of vessels, 229; procedures for implementing, 229; methods of, 230–31, 236, *passim;* RAN developed and underdeveloped fleets, 230; and non-RAN fleets, 230; same rules apply to all nations, 230; maximum total fleet established, 230; policies reduce fleet when necessary, 230; allowances made to increase underdeveloped national fleets, 230; licenses based on gear, carrying capacity, and age of vessel, 230; and issued to individuals, corporations, or nations, 230; and transfers, 230; types of fleet changes, 233–41 *passim;* license system for control of fleet size, 235–41 *passim;* reduction of license pool to control fleet size, 238–39; economic and resource management benefits, 241

Fleet development: methods of developing RAN fleets, 206–7; flag changes, 206–7, 212, 215–16; affected by catch allocations, 207; vessel transfers voluntary, 207; transfer advantages and disadvantages, 207; RAN requirements for transfers, 207; interregional transfers, 208; increased allocations as incentive for new fleet construction, 215; new vessel construction slow, 216

Flounder, 122–23

Food and Agriculture Organization, U.N., 173

France: IATTC member, 13; ICCAT member, 18; major harvester of principal market species of tuna, 26–27; effect of total allocation system on, 130; flag changes, 208; mentioned, 49, 64–65, 94, 108

Gabon, 18

Ghana, 18

Gilbert Islands, 174

Global management system: objectives of, 181; responsibilities, 181; functions, 181, 182; design for, 181–96 *passim;* work with national fisheries and related agencies, 183; structure of organization, 184; directorates for regional or separate functions, 184; enforcement, 184–85; allocation of responsibilities, 185–86; scientific research, 185; economic policies, 185; free exchange of information, 185; staffing, 186, 188–90; data collection on real time basis, 187–88; training research personnel, 188–89; membership, 190–91; voting systems, 192–93; action recommended to member nations, 192–93; advisory panels for major stocks or different species groups, 194; initiate global management by strengthening regional bodies, 195; optimum organization for world tuna management, 195–97; evolution from present regional organization as possible beginning, 195–97; coordinate activities of regional organizations, 196; establish central council, 196. *See also* Fisheries Management

Great Britain, 121–22

Greenland, 122

Guatemala, 27

Haddock, 122–23

Harvesting nations: on world basis, 11; RANs claim coastal water tunas, 26–27; six nations take 77 percent of world catch of tuna, 26–27; catch distribution international problem, 26–27; catch capabilities of RANs limited, 63

Hawaii, 176

Herring, 122–23

Historical participation in fisheries: rights to fish, 14; harvest of RANs, 63; U.S., Canada, and Chile tuna harvest, 103; nonregional participants, 112; length of time claims may be effective, 112; for fur seals, 121–22; for whales, 121–22; and northwest Atlantic fisheries, 121–22; justification for vessel replacements, 124; mentioned, 34, 125, 203

IATTC (Inter-American Tropical Tuna Commission): formation of, 13; member nations, 13; duties, 13–14; Commission's Yellowfin Regulatory Area (CYRA), 14; overall quota, 14; resource allocation conflict, 14–15; fleet growth problems, 15; enforcement, 17; collection of basic data, 23; staffing, 23, 187; funding, 23; management in eastern Pacific successful, 24; monitored RAN fleet vessels, 27; boundaries of RAN 200-mile zones determined, 28–29; management problems of fleet carrying capacity growth, 32; fleet catches in CYRA monitored, 36; fishing unrestricted west of CYRA, 36; management data confidential, 38–39; fisheries research and enforcement separated, 38–39; management objectives, 41, 57, 93; based on overall quota system, 53, 55; yellowfin regulatory program in CYRA, 53; present conservation program inadequate, 58; budget, 82; yellowfin management program, 93; overfishing prevented, 93; postclosure catch allowances, 114–15; resource allocations based on economic factors, 120; criteria for total allocation of yellowfin resource, 124–26; bait fish studies, 136; porpoise population estimates, 150; funding for porpoise research, 157; yellowfin allocation given for gear research, 164; recommendations for regulating porpoise fishing techniques, 170; porpoise kill quotas, 170; suggestions for management program, 170; international cooperation needed in tuna-porpoise problem, 170; structure as design for global organization, 183; real time data aquisition for eastern Pacific fishery, 188; training and technical assistance given member nations, 189–90; voting procedures, 192; open to nations other than original participants, 205; mentioned, 51, 68, 195

ICCAT (International Commission for the Conservation of Atlantic Tunas): formation of, 18; data collection problems, 18–19, 23, 187–88; size limits, 18; allocation, fleet size, enforcement problems, 19–20; competition between gear types, 19; staff and funding, 23, 187–88; structure as design for global organization, 183; training of fisheries scientists, 189–90; voting procedures, 192; advisory panels, 194; mentioned, 195

ICNAF (International Commission for Northwest Atlantic Fisheries): formation of, 122–23; multispecies fishery management, 122–23; voting procedures, 192; advisory panels, 194; mentioned, 124, 126, 127

Indian Ocean, 23, 59, 193, 197

Indian Ocean Fishery Commission. *See* IOFC

Indo-Pacific Fisheries Council. *See* IPFC

Inter-American Tropical Tuna Commission. *See* IATTC

International agreements, 120–21

International Commission for the Conservation of Atlantic Tunas. *See* ICCAT

International Commission for Northwest Atlantic Fisheries. *See* ICNAF

International fisheries management system: formation and structure of, 75–76, 91–92. *See also* Global management system

International organizations. *See* IATTC; ICCAT; ICNAF; International Pacific Salmon Fisheries Commission; International Pacific Halibut Commission; IOFC; IPFC; IWC

International Pacific Halibut Commission, 190

International Pacific Salmon Fisheries Commission, 190

International Whaling Commission. *See* IWC

IOFC (Indian Ocean Fishery Commission): formation of, 20; capabilities lacking, 20; exclusion of non-U.N. members, 21; staffing, 187–88; no responsibility for data collection, 187–88; mentioned, 195

IPFC (Indo-Pacific Fisheries Council): formation of, 20; capabilities lacking, 20; exclusion of non-U.N. members, 21; staffing, 187–88; not responsible for data collection, 187–88; mentioned, 195

Ivory Coast, 18

IWC (International Whaling Commission), 121–22, 192

Japan: tuna production and consumption, 11; IATTC member, 13; ICCAT member, 18; major harvester of tuna catch, 26–27; nonregional historical participant in eastern Pacific fishery, 112; northern fur seal catch, 121; effect of total allocation system on, 129–30; longline fishing, 154; aggregating devices to attract tuna, 156; market for, 174; billfish longline fishery, 174, 178; mentioned, 169

Korea, Republic of: ICCAT member, 18; major harvester of world catch of tuna, 26–27; longline fishing, 33–34, 154;

catches dropped, 33–34; effect of total allocation system on, 130

Latin America: coastal nations demand larger quota allocation, 53, 56, 61; fisheries jurisdiction to 200 miles by RANS, 54; RANs claim yellowfin catch shares, 1960, 125; special allowances granted nations in 1969 based on economic hardships, 125; mentioned, 48, 50, 63, 108, 111
Law of the Sea: U.N. conferences on, 22, 43, 54. *See also* U.N. Convention on the Territorial Sea and Contiguous Zone
Licenses: foreign flag vessels licensed by RANS, 43; alternative to maintaining exclusive fishing zones, 48–49; method of sharing foreign flag vessel catch, 48–49; sources of revenue from fees and catch, 48–49; method of maintaining resource access, 48–49; management system, 48–52 *passim;* established by Ecuador and Peru in 1940s, 54; single international system, 69, 73, 75; high fees set by RANS, 93; Ecuadorian fees high, 204; how issued, 230; who may buy, 230; transactions with fleet changes, 235–41 *passim;* transferable, 237–38; property value of, 237–38; bidding for, 237–38; charges levied for, 237–38; option to sell, 237–38; transactions handled through international management body, 238; fleet size reduced by limiting license pool, 238–39
Limited entry, 102
Longlining: basic technique, 9; by commercial and sport fishermen, 136; Korea, Japan, China, 154; tuna take low, 154; competition with purse seining, 205

Mackerel, 5
Marine Mammal Commission, 157
Marine Mammals Protection Act. *See* MMPA
Marlin, 173–75
Maximum net productivity (MNP), 147–53 *passim*
Maximum sustainable population (MSP), 147–53 *passim*
Maximum sustainable yield: objective of tuna management, 41, 93; billfish

fisheries, 177–78. *See also* Optimum sustainable yield
Mayaquez, Puerto Rico, 208
Medina safety panel, 141–42, 162
Mexico: IATTC member, 13; jurisdiction over fisheries, 22; yellowfin and skipjack catch areas, 29; fleet expansion, 45–46; license required to fish within 200-mile zone, 54–55; catch allocations under PAQ management system, 83; coalitions, 94–95; IATTC proposed allocation of yellowfin catch in CYRA, 126–27; tuna-porpoise associated fishery, 137; billfish longline fishery, 173–74; big game fishing, 176; prohibition of billfish fishing by Japanese in 200-mile zone, 178; fleet adequate to take low allocated quota, 199; mentioned, 27, 49, 66–68, 84, 94, 103, 112, 115, 167
Migrations. *See under individual species*
MMPA (Marine Mammals Protection Act): enactment of, 142; terms of, defined, 144–45; permits issued to take porpoise, 145; clarification of terminology of, 146–47; mandatory gear modification, 163–64; import sanctions provided against nations not protecting porpoise, 169; mentioned, 148, 150
MNP (maximum net productivity), 147–53 *passim*
MSP (maximum sustainable population), 147–53 *passim*
Multispecies fisheries: management in northwest Atlantic, 122–23; tuna-porpoise associated fishery, 136; complications, 205

National Marine Fisheries Service. *See* NMFS
National Science Foundation, 157
Nauru, 174
Netherlands, 122
Netherlands Antilles, 27, 112
New England, 122
New Guinea, 174
New Zealand: participant in eastern Pacific fishery, 27, 112; black marlin, 174; mentioned, 241
Nicaragua: IATTC member, 13; catch variations, 29; fleet expansion, 46; total allocation system curtails national fisheries development, 129; fleet size adequate to utilize allocation, 199; mentioned, 27, 66

NMFS (National Marine Fisheries Service): results of workshop for porpoise stock assessment, 145–54 *passim;* porpoise kill quota established, 145, 151; OSP concept of MMPA, 148; data obtained by aerial surveys and shipboard observer data, 149–51; proposed regulations for tuna-porpoise take, 151; experiments with fishing methods to reduce porpoise mortality, 157; porpoise-saving concepts in tuna fishing techniques, 158–61; modification of purse seine safety panel, 162–63; porpoise release apron, 162–63; food habit studies of tuna-porpoise schools, 171; mentioned, 169
Northern bluefin tuna. *See* Bluefin tuna
Northern fur seals, 120–21, 123
North Pacific Fur Seal Commission. *See* NPFSC
North Pacific Ocean, 122
Norway, 121–22
NPFSC (North Pacific Fur Seal Commission), 121, 190

Offshore spotted porpoise. *See* Spotted porpoise
Okhotsk, 121
Open-access concept: past application of, 14
Open-access fishery: defined, 54–55; curtailed in eastern Pacific, 54–55; PAQ management system, 58, 85; enables fleets to fish where fish concentrate, 68–69; essential to RANs with high quotas, 68–69; international licensing of all vessels, 68–69; unnecessary in 12-mile zone for international management, 72
Open-to-all regional management coalition, 113
Optimum population size: billfish fisheries, 177–78
Optimum sustainable population. *See* OSP
Optimum sustainable yield, 41, 144–45. *See also* Maximum sustainable yield
OSP (optimum sustainable population), 146–53 *passim*
Overfishing: yellowfin closure dates important, 198; mentioned, 95, 110. *See also* Fisheries management

Panama: IATTC member, 13; RAN in eastern Pacific yellowfin and skipjack fishery, 27; catch variations, 29; fleet

expansion, 46; catch allocations under PAQ management system, 83; large fleet for resources, 94, 104, 198–99; coalitions, 94; big game fishing, 176; mentioned, 49, 64–66, 82, 96, 114, 212
PAQ (partially allocated quota) management system: as international management plan and compromise of conflicting philosophies, 59, 93; methods of determining allocations, 62, 73–74; allocations as nontransferable guarantees of access, 62, 73–74; participant fee concept, 85, 223; open access to all fishing grounds, 85; partial allocation of overall catch, 85; international licensing, 85; management funded by fees, 85, 128; alternatives to, 85, 88–89; funds redistributed to participating nations, 85; yellowfin closure, 86–87; skipjack catch, 86–87; RAN fleet development, 88; RAN benefits, 81–90, 198, 202; procedures for fleet changes and licensing, 90–91; as open-to-all regional management coalition, 113; partial allocations based on resource adjacency, 120; total allocation system, 128; transferable property rights, 128; variability of fishable concentrations of tuna, 128; yellowfin and skipjack resources and management, 202–5; non-RAN benefits, 202–3; possible conflict between RANs and non-RANs over skipjack, 202–3; restriction of flag changes through international agreement, 209; effect on RAN fleet growth and total catch, 209; fleet economic problems, 209; methods for phasing in RAN allocations, 212; effect on fleet composition, 212, 214, 215; fleet growth evaluated by flag changes and by new construction, 212; transition phases based on ten-year allocation period, 212–14; three methods demonstrated, 212–14; skipjack quota possible, 214; transitional changes in fleet carrying capacity as catches increase, 214–15; catches of other tuna species, 222–23; measures to prevent decline of non-RAN vessel productivity with fleet transition, 222–23; fleet catches outside eastern Pacific, 222–23; unutilized RAN allocations available to non-RANs, 223; impact on fishery and fleet composition, 224–25; individual boat owner catch per capacity ton,

224–25; economic effect on non-RAN fleets, 225; methods for controlling fleet size, 229–41 *passim;* mentioned, 82, 96, 98, 105, 109, 116, 134, 135
Participant fee system: definition of, 50; method of compensating RANs for unutilized allocations, 62, 77–78; RAN benefits, 67; administered by single international system, 69; catch tax, requires monitoring system, 73–74; used for management funding, 73–74, 84–85; fisheries included under, 73–74; imposed on all harvesters, 73–74; disbursements to RANs, 76, 82; and average catches on high seas to harvesting nations, 77, 82; use of high seas portion of funds, 77–78; methods of collection and disbursement based on nontransferable allocations, 77–78, 84–85; alternatives to PAQ system, 84–85; under RAN coalition management, 96; disbursements and incentive payments, 223–24
Peru: claims 200-mile territorial seas, 22; catch variations, 29; fleet expansion, 46; skipjack fishery fluctuations, 46, 67; economic hardship on industry, 46; license fees, 49, 79; licensing effective within 200 miles, 54–55; yellowfin adjacency-based allocations, 63; big game fishing, 176; mentioned, 68, 94
Philippine Islands, 156
Pinas Bay, Panama, 176
Ponce, Puerto Rico, 208
Porpoise: avoid corkline, 141; avoid seiners, 149; population estimates, 150–51; techniques for saving, 158–61, 162, 164, 167–68; learning ability, 171–72; mentioned, 194–95. *See also individual species*
—mortality, fishery-induced, 136; annual estimates of NMFS observer data, 142–43; IATTC log book data, 142–43; increased with expansion of tuna fishery west of CYRA, 144; kill quotas, 145, 154; historical kill estimates, 150
Porpoise Rescue Foundation: established by U.S. tuna industry, 157
Portugal, 18
Postclosure fishing: effect of, on fishery, 116–17

Principal market species: definition of, 5; six nations harvest 77 percent of world catch, 26; half of catch taken within 200 miles of shore, 26; U.S. as major harvester, 26–27; catch tonnage and fleet carrying capacity, 32
Puerto Rico, 208
Purse seining: basic technique, 9–10; tuna-porpoise associated fishery, 136–37; only economically efficient gear to harvest tuna, 154–55; design and modification for tuna-porpoise fishing, 154–68 *passim;* mandatory modification to include safety panel and porpoise release apron, 163–64; competition with longlining, 205

Quotas: overall or global, 14; set by IATTC on yellowfin, 53; differences among participating nations on catch allocations, 53; special allocations based on economic hardship allowed RANs, 54; yellowfin special allocation increasing, 54; existing management system inadequate for eastern Pacific, 57. *See also* PAQ management system

RANs (resource adjacent nations): want preferential rights to harvest tuna in their coastal waters, 27, 43–44, 200; philosophical conflicts in fishery management, 42; control of coastal zones to 200 miles seaward, 43; agreements with foreign flag vessels, 43; license foreign vessels, 43–44, 48; catch quotas of yellowfin tuna estimated in CYRA and beyond, 43–44, 200; fishing restrictions affect non-RAN catches, 44; vessels inadequate to harvest tuna in their coastal waters, 45; tuna migrations variable, 45; effect of exclusive national fisheries, 46–47; fleets of, 47, 88, 94; licensing only other RAN vessels, 48; preservation of resource access in 200-mile zone, 48; maximize catches and revenues, 48; limit on licenses sold, 48; revenue from license fees, 51; tuna catch rates, 89; coalition management system, 94, 100–8 *passim;* coalition with non-RANs, 114–16
Regional management system, 108–13
Richey, Judge Charles R., 146

Sailfish *(Istiophorus platyperus)*, 173–74
Satellite technology: proposed to determine positions of vessels at sea, 37
Scombridae, 173
Seals, northern fur, 120–21, 123
Secondary market species: migrations, 5; definition of, 5; world catch, 5, 6
Senegal: ICCAT member, 18; harvesting nation in eastern Pacific tuna fishery, 27, 112
Simulated annual catches, 99
Simulation models: for yellowfin, 77; for yellowfin, skipjack, and combined catches, 97–98
Skipjack tuna *(Katsuwonus pelamis):* migrations, 5, 45–46, 204; world catch, 6; catch by ocean, 7; catch in eastern Pacific, 45, 56, 144, 203–5, 223; distribution in eastern Pacific, 45, 67, 203–4; economic hardship when migrations vary, 46; loss in catch with RANS closing 200-mile zone, 46; no quotas, maximum catch encouraged, 61; management, 61; management under PAQ system, 86–87, 202–3, 214; catch quota possibility, 88; simulated projections of catch, 98; Chilean resources, 108; not fully utilized, 128; closure methods should parallel yellowfin, 201; allocations under PAQ management, 202, 215, 226, 227; participant fees, 202, 215; open access all year, 202–3, 215; fishing intensity high after yellowfin closure, 202; open access fishery as economic hardship for Ecuador, 203–4; harvesting ability, 203–4; mentioned, 46–47, 118, 194
South Africa, 18
Southern bluefin tuna. See Bluefin tuna
Sovereign rights: 12-mile territorial sea, 69–72
Spain: tuna production, 11; ICCAT member, 18; major harvester of tuna, 26–27; participant in eastern Pacific fishery, 27, 112; effect of total allocation system on, 130
Spawner-recruit (S-R), 97
Spearfish, 174
Spinner porpoise *(Stenella longirostris),* 137, 168
Sport fishing: tuna, 136; billfish, 172–73; conflict with commercial fishermen, 172–73, 205

Spotted porpoise *(Stenella attenuata),* 137; offshore range of stock, 138, 150–51, 168; kill quota set, 152
Stock-recruitment relationship, 98
Striped marlin *(Tetrapturus audax),* 173–74
Striped porpoise *(Stenella coeruleoalba),* 137
Super seiners, 138
Swordfish *(Xiphias gladius),* 173–74

Territorial sea, 72
Total allocation system: high degree of management control, 123; participating nations, 123, 124; encourage fleet development, 124; criteria for allocation, 124; participant right to allocate catch, 124; new-entrant nation a major obstacle to management, 127–28; transferability of quotas, 127–28; reduced competition, effect on technological advances, 128–29; discourages fleet development, 128–29
Transferable property rights. *See* Allocations
Tuna Conventions Act, 17–18
Tuna Foundation, 157–58, 162
Tunas: migrations, 5; species of, 5; world catches, 6–9; development of world fishery, 6–8; management problems, 22; multispecies fishery, 24; study of species complexes needed, 24; resources limited, 25–26; porpoise-associated schooling and resulting problems, 25, 138, 168–72; resource subject to rapid changes, 49; mobility of, 49; variations in distribution and abundance, 66. *See also individual species*
Two-hundred-mile zone. See Zones of economic influence

U.N. Convention on the Territorial Sea and Contiguous Zone (1958), 28–29, 40–41. See also Law of the Sea
United States: tuna consumption, 11; tuna production, 11; IATTC member, 13; Tuna Conventions Act, 17–18; ICCAT member, 18; jurisdiction claims, 22; porpoise kill associated with yellowfin tuna, 25; major harvester of the world catch of tuna, 26–27; allocations, 53, 63–81, 83, 125–27, 130, 201; legis-

lated insurance fund to compensate owners of vessels seized by RANs, 54; basis for allocating RAN quotas, 61; historic catches, 61; yellowfin catch, 63, 125–27; PAQ management regulations, 83; effect of exclusion from fishing in RAN 200-mile zone, 100; historical participation in fishery, 103, 112; purseseine tuna fleet, 107; dominates eastern Pacific tuna fishery, 108; cooperation in fisheries management, 108; research funding and enforcement monitoring, 108; coalition participation, 108–10; fleet flag changes, 109–10; fleet fish eastern Atlantic after CYRA closure, 111; and fish western Pacific, 111; northern fur seals, 121; IATTC recommended yellowfin catch in CYRA, 125–27; vessel carrying capacity, 125–27; competitive bidding, 131; legislation to stop fishery-induced porpoise mortality, 136; technological aspects of the tuna-porpoise fishery, 136; U.S. tuna seiners required to have Medina safety panels, 142; basis for establishing porpoise kill quotas, 147; tuna fleet leader in developing techniques to minimize porpoise mortality, 152, 156–57, 167; tuna-porpoise problem, 162–68 *passim;* gear modifications mandatory for vessels licensed to fish on porpoise, 163–64; import sanctions placed against nations not protecting porpoise, 169–70; billfishes, marlin, sailfish, 173–74, 178; benefits to U.S. from transfer of RAN allocations, 201; extension of open fishing period, 201; bidding for RAN allocations, 201; fleet change from one ocean to another, 206; flag changes, 207; opposition to flag changes, 208–9; tuna exploration in western Pacific, 241; mentioned, 27, 32, 50, 56, 79, 84, 94, 114, 116, 173, 192, 199

U.S. National Forests: timber harvested by competitive bids, 131

USSR: ICCAT member, 18; longline and purse seine fleets under construction, 110; possible mothership operation in eastern Pacific tuna fishery, 110; northern fur seals, 121

U.S. tuna fishing industry: porpoise-saving concepts, 167

Venezuela: participant in eastern Pacific fishery, 27, 112; white marlin, 173–74; sailfish, 173–74

Vessels: designed to catch tuna and tuna-like fish, 33; harvest of secondary market species, 33; historical participation in fishery, 34–35; new entrants, 34–35; underdeveloped fleets, 34–35; flag changes, 206–9 *passim;* U.S. vessel transfers from individual to processor ownership, 224; control of entry into fishing fleet, 229–41 *passim*

Western Europe, 169; swordfish market, 174

Whale resources: total catch allocation at international level, 120–23

Whitebelly spinner porpoise, 137; range of stock in eastern Pacific, 140; population estimate, 150–51; U.S. kill quota set, 152

Yellowfin tuna *(Thunnus albacares):* migrations, 5; world catch, 6; catch by ocean, 7; eastern Pacific fishery, 25; tuna-porpoise associated fishery, 25; fishery regulated by catch quota in eastern Pacific, 35; catch limited by size in Atlantic, 35; fishery regulated by IATTC in CYRA, 36; fishing west of CYRA unrestricted, 36; mixing rate between offshore and coastal waters, 44; catches in eastern Pacific, 44, 63, 74, 98; tagging data, 44; spawner-recruitment relationship, 44–45; effect on catch if coastal waters closed by RANs, 44, 46; substantial portion of catch based on adjacency allocations, 56; historical fishing rights vs. adjacency resource rights, 60, 203; U.S. percent of catch, 63; operation of fishery under participant fee system, 78–79; closure of fishery, 86; overall quota, 86; monitoring catches of vessels at sea, 86; RAN allocations under PAQ management system, 87; spawner-recruit relationship to increased catch beyond 200 miles, 97; recruitment based on three simulated model situations, 98; ten-year catch projection using simulated model data, 98; tonnage taken inside vs. outside 200-mile zone, 98; overfishing possibility without management, 104; overall quotas set, 144; fishery expanded west of CYRA, 144; closure methods, 199–201 *passim;* overall quota under PAQ management system, 202; guaranteed access for RANs, 202; portion of participant fees given to RANs, 202; harvesting ability, 203; mentioned, 116, 194, 198

Yucatan Peninsula, 174

Zero mortality, 145
Zones of economic influence, 43